Marine Genetic Resources, Access and Benefit Sharing

T0227568

Access to genetic resources and Benefit Sharing (ABS) has been promoted under the Convention on Biological Diversity, with the aim of combining biodiversity conservation goals with economic development. However, as this book shows, since its inception in 1992, implementation has encountered multiple challenges and obstacles.

This is particularly so in the marine environment, where interest in genetic resources for pharmaceuticals and nutrients has increased. This is partly because of the lack of clarity of terminology, but also because of the terms of the comprehensive law of the sea (UNCLOS) and transboundary issues of delineating ownership of marine resources.

The author explains and compares relevant provisions and concepts under ABS and the law of the sea taking access, benefit sharing, monitoring, compliance, and dispute settlement into consideration. He also provides an overview of the implementation status of ABS-relevant measures in user states and identifies successful ABS transactions. A key unique feature of the book is to illustrate how biological databases can serve as the central scientific infrastructure to implement the global multilateral benefit sharing mechanism, proposed by the Nagoya Protocol.

Bevis Fedder is a Postdoctoral Research Associate at the University of Bremen, Germany.

Marine Genetic Resources, Access and Benefit Sharing

Legal and biological perspectives

Bevis Fedder

 Routledge
Taylor & Francis Group

LONDON AND NEW YORK

 from Routledge

First edition published 2013
by Routledge

2 Park Square, Milton Park, Abingdon, Oxfordshire OX14 4RN
711 Third Avenue, New York, NY 10017

Routledge is an imprint of the Taylor & Francis Group, an informa business

First issued in paperback 2017

British Library Cataloguing in Publication Data
A catalogue record for this book is available from the British Library

Library of Congress Cataloging-in-Publication Data
Fedder, Bevis.
   Marine genetic resources, access, and benefits sharing : legal and
   biological perspectives / Bevis Fedder.
        pages cm
   Includes bibliographical references and index.
   1. Marine resources conservation – Law and legislation.
   2. Marine biodiversity conservation – Law and legislation.
   3. Biodiversity conservation – Law and legislation. 4. Convention on
   Biological Diversity (1992). Protocols, etc., 2010 Oct. 29. I. Title.
   K3485.F43 2013
   346.04´695616-dc23   2012048241

ISBN13: 978-0-415-83055-3 (hbk)
ISBN13: 978-1-138-57315-4 (pbk)

Typeset in Times
by HWA Text and Data Management, London

Contents

List of illustrations   vii
Preface and acknowledgements   ix
List of acronyms and abbreviations   x

1 **Introduction**   1
Problems and objectives 1

2 **Factual background**   3
Introduction 3
Marine biotechnology 4
Distribution of marine species 16
Conclusions 17

3 **Access and benefit sharing in the marine realm**   18
Introduction 18
Sovereign rights and the common heritage of mankind 26
Sovereign rights over genetic resources 34
International instruments for activities on marine genetic
   resources 42
National regimes on management of marine genetic resources 77
Selected case studies 106
Conclusions 110

4 **Weak points of the access and benefit sharing regime**   112
Introduction 112
Injustice 112
Ineffectiveness 118
Hampering research and development 119
Conclusions 120

5 Biological databases for improving the ABS system   **122**
Introduction 122
Bioinformatics 122
Tracing products to their source countries and monitoring users 124
Exemplary application of selected databases 155
Conclusions 175

6 Conclusions   **177**

Notes   180
References   183
Index   211

Illustrations

Figures

3.1   Maritime zones   24
5.1   Disciplines contributing to bioinformatics   123
5.2   Sketch of the search window of the Dictionary of Marine Natural
      Products (CD-ROM version) using spongothymidine as an example   129
5.3   Sketch of the RÖMPP search pane using ecteinascidin as an
      example   131
5.4   Sketch of the search screen interface of the Merck Index   134
5.5   Map of an exemplary species distribution in OBIS, using the cone
      snail, *Conus magus*, as an example   150
5.6   Map of a GBIF record, using *Conus magus* as an example   153

Tables

2.1   Non-exhaustive enumeration of marine natural compounds and
      their (commercial) derivatives   14
3.1   ABS-relevant elements within international agreements   78
3.2   Categorization of ABS-relevant legislation of Australia, the
      Northern Territory and Queensland   96
5.1   Biological databases analysed in this chapter   124
5.2   Summary of the databases analysed for their content, application,
      legal form, rules of feeding and retrieving data, data limitations
      and ABS elements   156
5.3   Linking biological molecules to their source species using selected
      databases   158
5.4   Linking gene products to their source species using the database
      GenBank   160
5.5   Linking the source species to their source states using the GBIF
      geographical database   162
5.6   Non-exhaustive list of research papers, patents, brands, firms and
      cross-references which databases contain for biological molecules   164

Boxes

5.1 Selected elements of an entry from the Dictionary of Marine
    Natural Products, using spongothymidine as an example   128
5.2 Elements of an entry from RÖMPP Online using ecteinascidin as
    an example   130
5.3 Elements of a monograph from the Merck Index using
    ecteinascidin as an example   133
5.4 Elements of a record from PubChem Compound using
    ecteinascidin as an example   139
5.5 Elements of the entry on the growth hormone construct
    opAFP-GHc2   141
5.6 Elements of a UniProt entry using Deep Vent DNA Polymerase as
    an example   145
5.7 GBIF countries with occurrence records of *C. magus*   154

Preface and acknowledgements

This work was funded by and conducted as part of the Bremen International Graduate School for Marine Sciences, 'GLOMAR'. GLOMAR is funded by the German Research Foundation under the Excellence Initiative by the German federal and state governments. The book represents the culmination of my work from 2009 to 2012 and has been successfully submitted as a doctoral thesis to the Faculty of Law at the University of Bremen. The work has been updated to include latest references and developments and was finalized in December 2012.

It is a pleasure to thank many different people who made this work possible. I am particularly grateful to my supervisors, Prof. Dr. Gerd Winter and Prof. Dr. Antje Boetius, whose support, invaluable advice, trust and encouragement provided the most important input to this work. I couldn't think of any better supervisors!

I owe further gratitude to Prof. Dr. Sabine Schlacke, Prof. Dr. Josef Falke, Dr. Till Markus, Dr. Evanson Chege Kamau, Dr. Olaf Dilling and Anja Strüve, all of whom have provided useful comments, fruitful discussions and support during elaboration of this work and the thesis colloquium. Special thanks go to Dr. Till Markus, whose advice on legal issues did not only improve this work substantially, but his friendship and attitude towards life have inspired me greatly.

Many thanks go to my colleagues at the Research Center for European Environmental Law and GLOMAR, particularly Prof. Dr. Dierk Hebbeln, Dr. Uta Brathauer, Dr. Christina Klose, Jutta Bülten, Carmen Murken and Antje Spalink.

I want to thank everyone from my scientific network for having assisted me and thereby contributed to this book.

I owe special gratitude to my friends, Nina Maier, Lisa Marquardt, Bryna Flaim, Ingo Unterweger and Volker Prott, and of course to my family, particularly both of my parents. Their love and support was the greatest stimulus to embark upon this work.

Acronyms and abbreviations

ABS	Access and Benefit Sharing
ACTS	African Centre for Technology Studies
AJSTD	Asian Journal for Science and Technology Development
ASCII	American Standard Code for Information Interchange
ASEAN	Association of South East Asian Nations
AUS	Australia
BfN	Bundesamt für Naturschutz (Federal Agency for Nature Conservation)
BLAST	Basic Logical Alignment Search Tool
BMC	Biomed Central
CABI	Commonwealth Agricultural Bureau International
CAMERA	Community Cyberinfrastructure for Advanced Marine Microbial Ecology Research and Analysis
CAS	Chemical Abstracts Service
CBD	Convention on Biological Diversity
CE	Chief Executive
CEO	Chief Executive Officer
CHM	Clearing-House Mechanism
CISDL	Centre for International Sustainable Development Law
CNA	Competent National Authority
COP	Conference of the Parties
CRC	Chemical Rubber Company
CS	Continental Shelf
DBE	Department of Business and Employment (Northern Territory)
DDT	Drug Discoveries and Therapeutics
DEWHA	Department of the Environment, Water, Heritage and the Arts (Australia)
DFG	Deutsche Forschungsgemeinschaft (German Research Foundation)
DHA	Docosahexaenoic acid
DMNP	Dictionary of Marine Natural Products
DNA	Deoxyribonucleic acid
DoR	Department of Resources (Northern Territory)

DSDI	Department of State Development and Innovation (Queensland)
EC	European Community
EEZ	Exclusive Economic Zone
EPA	Environment Protection Agency (Queensland)
EP-ABS	Panel of Experts on Access and Benefit-Sharing
EU	European Union
FAO	Food and Agriculture Organization of the United Nations
FDA	Food and Drug Administration
FTP	File Transfer Protocol
GBIF	Global Biodiversity Information Facility
GEF	Global Environment Facility
GFP	Green Fluorescent Protein
GIS	Geographical Information System
GR	Genetic Resource
GSK	GlaxoSmithKline
GTLE	Group of Technical and Legal Experts
HIV	Human Immunodeficiency Virus
ICIMOD	International Centre for Integrated Mountain Development
ICNP	Intergovernmental Committee for the Nagoya Protocol on Access to Genetic Resources and the Fair and Equitable Sharing of Benefits Arising from their Utilization
ICJ	International Court of Justice
ILC	International Law Commission
ILM	International Legal Materials
ILR	International Law Reports
IOC	Intergovernmental Oceanographic Commission
IODE	International Oceanographic Data and Information Exchange
IPR	Intellectual Property Rights
ISA	International Seabed Authority
ITLOS	International Tribunal for the Law of the Sea
ITPGRFA	International Treaty for Plant Genetic Resources for Food and Agriculture
IUCN	International Union for Conservation of Nature
IUPAC	International Union of Pure and Applied Chemistry
JCVI	J. Craig Venter Institute
KGS	Kansas Geological Survey
MAT	Mutually Agreed Terms
MqJICEL	Macquarie Journal of International and Comparative Environmental Law
MTA	Material Transfer Agreement
NAPRALERT	Natural Products Alert
NCBI	National Center for Biotechnology Information
NIH	National Institutes of Health
NMNH	National Museum of Natural History
OBIS	Ocean Biogeographic Information System

OECD	Organisation for Economic Co-operation and Development
PIC	Prior Informed Consent
PLoS	Public Library of Science
PU	Penalty Unit
R&D	Research and Development
RAFI	Rural Advancement Foundation International
RECIEL	Review of European Community and International Environmental Law
RNA	Ribonucleic acid
SBSTTA	Subsidiary Body on Scientific, Technical, and Technological Advice
SMILES	Simplified Molecular Input Line Entry Specification
TIBTECH	Trends in Biotechnology
TK	Traditional Knowledge
TRIPS	Trade-Related Aspects of Intellectual Property Rights
UN	United Nations
UNCLOS	United Nations Convention on the Law of the Sea
UNCTAD	United Nations Conference on Trade and Development
UNDOALOS	United Nations Division for Ocean Affairs and the Law of the Sea
UNEP	United Nations Environment Programme
UNESCO	United Nations Educational, Scientific and Cultural Organization
UNGA	United Nations General Assembly
UniProt	Universal Protein Resource
UNU-IAS	United Nations University – Institute of Advanced Studies
WCMC	World Conservation Monitoring Centre
WG-ABS	Ad Hoc Open-Ended Working Group on Access and Benefit-sharing
WTO	World Trade Organization

1   Introduction

Genetic resources[1] have been a hot topic in recent decades. The utilization of genetic resources creates benefits, and many stakeholders are keen to secure their share of any benefits. As these stakeholder groupings often have divergent interests, this could lead to conflict, so the legal regulation of activities on genetic resources is imperative. The key international instrument for regulating such activities is the access and benefit sharing (ABS) regime under the Convention on Biological Diversity (CBD) (UNGA, 2005, preamble 2). Since its inception, the ABS regime has been criticized on many grounds. Consequently, it has evolved, culminating in the adoption of the Nagoya Protocol in 2010. However, the ABS regime is not the only international instrument that regulates activities on genetic resources. With regard to the marine environment, the United Nations Convention on the Law of the Sea (UNCLOS) is the key international instrument regulating all activities. This work aims to analyse both instruments, identify and assess some of the major problems in these regimes and provide solutions to these problems.

Problems and objectives

Although ABS has constantly evolved over the past decades, it remains a technically challenging and legally complex issue (Pisupati, 2008, p.1). First, pharmaceutical research on genetic resources experienced a depression in the mid-2000s (Moran *et al.*, 2001, p.508; Butler, 2004, p.2141). As the pharma-industry is one of the major users of genetic resources, this depression has led experts to conclude that ABS may eventually become obsolete (ten Kate and Laird, 2000a, p.261).

Second, key concepts of the ABS regime have remained ambiguous. This includes: a) the relevance of sovereignty, sovereign rights and the common heritage of mankind principle for defining the legal status of genetic resources and b) the definition of 'genetic resources' and related concepts, such as 'utilization' and 'access' (Young, 2008, p.119).

Third, activities on marine genetic resources are not only regulated by the ABS regime under the CBD but also by several additional sub-regimes under UNCLOS. This raises questions about applicable provisions and the legal relationship between the different regimes. In addition, it is also important to examine how member states have solved the dual applicability of both conventions within their domestic legislation.

Fourth, the current ABS system stipulates bilateral exchange, under which a user of marine genetic resources returns a share of the benefits of utilization to the provider state where the user accessed the resource. This system has three major problems (Winter, 2009, pp.19 and 25–27):

- It is unjust because it fails to include other source states sharing the same marine genetic resource when stipulating benefit sharing.
- Second, it is ineffective, because provider states cannot unilaterally control the whole value chain of a particular marine genetic resource after exportation (downstream control).
- Third, because of ineffective downstream control, provider states can be prompted to adopt overly stringent access legislation, stifling research and development ('research chill'). This would take away the very foundation on which ABS is based.

In light of what has been described above, this work is structured into the following chapters. Chapter 2 provides the scientific background by illustrating the important role that marine genetic resources continue to play in product development. Chapter 3 analyses and compares the current international instruments governing activities involving marine genetic resources. Moreover, this chapter: a) strives to explain the key debated concepts relating to ABS, b) examines the implementation of international regimes by a range of typical user states and c) closes with selected case studies on ABS. Chapter 4 explains in detail why current conduct in ABS transactions is unjust, ineffective and may in fact be hampering research and development. In response, Chapter 5 introduces and applies biological databases as instruments that could promote justice and effectiveness and support research and development. Chapter 6 concludes the work.

2　Factual background

This chapter outlines the scientific background and provides examples of products derived from the utilization of marine genetic resources.

Introduction

The oceans are a rich source of biological molecules which could be useful for research and development (Bollmann *et al.*, 2010, pp.176–195). This is a result of the high species richness and the multitude of ecological influences on marine organisms. The high species richness of the oceans depends primarily on the size and complexity of the marine environment. The surface of the Earth is covered up to 70.8 per cent (362×10^6 km^2) by the oceans, which accounts for 99 per cent (1.370×10^7 km^3) of the volume that is known to sustain life (de Fontaubert *et al.*, 1996, p.5; WCMC, 1996, p.3; UNEP, 2006, p.v). Within the oceans, a multitude of defined habitats can be found. These habitats comprise: complex physical structures at the coast such as reefs, kelp forests, mangrove forests and rocky shores; stratified layers of water masses throughout the whole water column and various micro-habitats in the deep sea, such as hydrothermal vents, cold seeps, seamounts, cold water corals, sponge gardens and even carcass droppings from shallower depths (Grassle, 1989, p.13; Briggs, 1994, p.132; May, 1994, p.108). The high variety and complexity of marine habitats have facilitated the speciation of a highly diverse marine biota (Williamson, 1997, pp.12; Thorne-Miller, 1999, p.50). For example, species richness for coral reefs alone has been estimated at approximately 950,000 species globally (Reaka-Kudla, 1997, p.93). Species richness of deep-sea sediments has been extrapolated at 10 million species (Grassle and Maciolek, 1992, p.333). Global marine meiofauna is predicted to represent 100 million species (Lambshead, 1993, p.11),[1] and marine microbial diversity is even estimated to comprise 1 billion species (Amaral-Zettler *et al.*, 2010, p.242). Currently, only about 250,000 marine species have been described (Groombridge and Jenkins, 2002, p.122).

Marine organisms are exposed to various ecological influences. These include abiotic influences and interaction with other organisms. Abiotic influences include: light intensity; pressure; temperature and the chemical composition of the water, which affects pH, toxicity and salinity as well as oxygen and nutrient concentration. Interactions with other organisms include: competition for space, light and nutrients; predation; defence from predation and attachment; facilitation of reproduction and communication (Kornprobst, 2010, pp.13–24). As a response, marine organisms have adapted and evolved complex systems of biochemical reactions. Nucleic acids constitute the basis of these biochemical reactions or, more specifically, genes, which are inherited from generation to generation. The expression of information stored within genes produces proteins, carbohydrates, lipids and a plethora of other biological molecules, all of which ensure survival and propagation in the marine environment. They are often species-specific[2] and highly potent in small concentrations because they dilute quickly in water (Williams *et al.*, 1989, p.1190; Scheuer, 1990, p.173; Shimizu, 2000, p.30; Harper *et al.*, 2001, p.4; Füllbeck *et al.*, 2006, p.348).

These classes of biological molecules are the raw materials for the utilization of (marine) genetic resources. As their mode of biological action closely resembles processes applied in biotechnology, marine biological molecules carry great potential for various fields of application.

Marine biotechnology

The deliberate utilization of biomolecules for research and development falls roughly under the category of biotechnology. The Organisation for Economic Co-operation and Development (OECD) provides the most specific definition of biotechnology, describing it as the 'application of science and technology to marine living organisms, as well as parts, products, and models thereof, to alter living or non-living materials for the production of knowledge, goods, and services' (OECD, 2012). This definition is still deliberately broad and encompasses a vast set of techniques involved in the manipulation of biomolecules.

Marine biotechnology is a young subset of biotechnology. The special physical characteristics of seawater (higher pressure, less light and lower temperatures with increasing depth) made it hard for scientists to access and adequately sample the oceans (Colwell, 2002, p.218). With the advancement of modern scuba diving technologies, large parts of the oceans and their organisms have become increasingly accessible for biotechnology. In addition, marine biotechnology has also benefited from general biotechnological advances, such as high throughput screening, which increases the efficiency of research methods (Bugni *et al.*, 2008, p.1095). The products derived from marine biotechnology are manifold and cover several distinct fields with both commercial and non-commercial applications.[3] The overall commercial value of products amounts to as much as US$2.2 billion (Pisupati *et al.*, 2008, p.52) (see also Table 2.1 at the end of this section for selected products).

Agrochemicals

Insect pests, weeds and phytopathogenic fungi cause substantial harm to agricultural crops. Agrochemicals or pesticides are the major tool for protecting crops from these damaging pests. In 2001, world pesticide sales amounted to almost US$33 billion: 44 per cent for herbicides, 28 per cent for insecticides, 19 per cent for fungicides and 9 per cent against other pests (rodenticides, molluscicides and fish/bird pesticides) (Kiely *et al.*, 2004, p.4).

Most of these chemicals are synthetically derived. However, growing concerns about the impact of synthetic chemicals on human and environmental health, as well as the increasing resistance of pests to such pesticides, have led to a shift in research towards naturally-derived protection (Dayan *et al.*, 2009, p.4022). Currently, only about 30 'natural' agrochemicals exist, and most of these are terrestrially derived (Copping, 2001). However, marine-derived biomolecules have also great potential as agrochemicals. This applies particularly to insecticides, because insects are almost exclusively terrestrial and freshwater animals and have, as such, evolved little resistance to chemicals derived from marine organisms (Peng *et al.*, 2003, p.2251). In fact, the only commercial marine-derived agrochemicals nereistoxin and its analogues – thiocyclam, bensultap and cartap – are insecticides (Llewellyn and Burnell, 2000, p.64). Nereistoxin was originally isolated from the marine annelid *Lumbriconereis heteropoda*, which occurs in the Red Sea (Sattelle *et al.*, 1985, p.38). The analogue thiocyclam is sold under the trade name Evisect S®, and is marketed jointly by Arysta LifeScience Corporation and Nippon Kayaku. Global annual sales in 2005 yielded US$7 million (Dewar, 2005, pp.62 and 71).

As well as insecticides, biomolecules produced by marine organisms have great potential as herbicides. In particular, coral reefs, if they are healthy, are highly resistant against overgrowth by algae, indicating the presence of molecules within corals that deter the attachment of algae (Llewellyn and Burnell, 2000, p.64).

Antifoulants

Any unprotected, submerged surface, such as a ship's hull, will become fouled by the overgrowth of marine organisms. Fouling is beneficial insofar as organisms can use automatic water movement for feeding and waste removal (Rittschof, 2001, p.545). Fouling may be so dense that the increased friction between seawater and hull results in a loss of speed, manoeuvrability and an increase of fuel consumption of up to 50 per cent. These costs result in an annual expenditure for the shipping industry exceeding US$6.5 billion (de Nys and Steinberg, 2002, p.244; Bhadury and Wright, 2004, p.563). Early efforts to reduce fouling resulted in the development of antifouling paints that incorporate biocidal compounds (Howell and Evans, 2009, p.203). When an organism tries to settle on a hull, these biocides kill the organism. One of the most successful antifoulant paints was tributyl tin. However, its high toxicity and the low degradability leading to bioaccumulation and biomagnification (Meng *et al.*, 2005, p.141), resulted in a ban of tributyl tin and similar toxic compounds in 2003 according to Article 4 of the International Convention on the Control of Harmful Anti-Fouling Systems on Ships. The search for effective and yet environmentally sound alternatives has led to the development of natural antifoulants. Natural antifoulants have several advantages over conventional antifoulants: they are less toxic, effective at low concentrations and biodegradable. They also have a broad spectrum of antifouling activity and their ecological effects are reversible. Currently, over 145 such antifoulants have been identified and isolated from marine sources, but only three, Sea-Nine 211™, Netsafe® and Pearlsafe®, have reached the market (Raveendran and Limna Mol, 2009, p.508). The latter two products, produced by Wattyl Australia, have been derived from a suite of brominated furanones produced by the red seaweed, *Delisea pulchra*, occurring in waters around Australia, East Asia and the Antarctic (Dworjanyn *et al.*, 2006, p.154; Sims, 2000, p.43). The active molecule of Sea-Nine 211™ is isothiazolon, which is produced by a Caribbean soft coral, *Eunicea* sp. Sea-Nine 211 is highly effective against fouling from bacteria, algae and barnacles and is marketed by Rohm and Haas, a subsidiary of the Dow Chemical Company (Jacobson and Willingham, 2000, p.109; Raveendran and Limna Mol, 2009, p.512).

An alternative to the toxic paints discussed above are 'living paints' containing bacteria which produce bioactive molecules that inhibit fouling. Such products are still at an early stage in research and development (de Nys and Steinberg, 2002, p.245).

Bioremediation

The anthropogenic release of hydrocarbons (crude oil, petroleum) and the discharge of industrial and domestic waste into rivers and the sea raise concerns about the health of the marine environment and the long-term impact on humans (Ali and Llewellyn, 2009, p.572). One approach to remove such toxic substances is bioremediation. Bioremediation involves using micro-organisms to act as pollution control agents by detoxifying and removing heavy metals and hydrocarbons from the aqueous environment (Banat *et al.*, 2000, pp.495–504; UN, 2007, para 168).

For removing hydrocarbons from aqueous environments, microbially-produced surfactant molecules contain hydrophilic as well as hydrophobic structures which emulsify hydrocarbons. The resulting dispersion increases the area available for microbial colonization and the natural degradation of hydrocarbons (Head *et al.*, 2006, p.173). Microbially-produced surfactants are more environmentally-friendly because of their lower toxicity, high biodegradability and high, specific activity even at environmental extremes (temperature, pH, salinity) (Desai and Banat, 1997, p.47).

Various marine microorganisms are subject to scientific study (Cohen, 2002, p.189; Satpute *et al.*, 2010, p.436). The bacterium *Acinetobacter calcoaceticus*, isolated from the Mediterranean Sea, is the producer of a heteropolysaccharide protein which was exploited commercially as an emulsion between 1975 and 1990 by Petroferm USA (Rosenberg and Ron, 2001, p.93). The emulsions were used for cleaning oil sludge from ballast water tanks. Another ubiquitous bacterium, *Pseudomonas aeruginosa*, is the biological source of rhamnolipids, which had already been successfully used for treating Alaskan shorelines during the Exxon Valdez oil spill in 1989 (Bragg *et al.*, 1994, p.413; Prince, 1997, p.158; Kimata *et al.*, 2004, p.41). Rhamnolipids are the source of a series of biosurfactant products marketed by Jeneil Biosurfactant for bioremediation, including JBR425. These rhamnolipids also find application as pesticides, e.g., under the brand Zonix™ (Mulligan, 2009, p.372; Kaczorek *et al.*, 2010, p.364).

Regarding the removal of heavy metals – primarily lead, cobalt and cadmium – surfactant molecules must be able to form metal complexes that are soluble and environmentally stable. Although research is still at an early stage, various marine organisms have been already found that either produce such compounds or absorb heavy metals from the environment (Das *et al.*, 2009, p.4887).

Biofuels

The increasing anthropogenic concentrations of carbon dioxide and other greenhouse gases in the atmosphere, which enhance global warming, have led to the search of alternative, carbon-neutral sources of fuel. Palm oil and other oleiferous crops grown in monocultures are one source of biofuel. On the negative side, such monocultures cause harmful environmental and social consequences (loss of biodiversity, destruction of natural sinks, soil degradation, water scarcity, competition with food production), thereby compromising the positive effects of biofuels (German Advisory Council on Global Change, 2010, pp.1 and 19). Some alternatives to agriculturally grown biofuels are microalgal biofuels and marine microbial fuel cells.

Culturing microalgae for biofuel production, in contrast to monocultures, entails various advantages: a) microalgae have fast growth rates and can be harvested at short intervals; b) culturing microalgae does not require arable land and thus does not interfere with food security; c) microalgal cultures can use industrial flue gas as a carbon source and d) the consumption of algal biofuels produces fewer emissions than fossil fuels (Mutanda *et al.*, 2011, p.57). Although microalgae occur in freshwater and marine environments, it is mainly marine species belonging to diatoms that have attracted interest owing to their very high oil contents that can be easily transformed into biofuel. However, given the novelty of algal biofuels, there has been little progress in the research and development of commercial biofuels. Major issues concern the costs of production: €50 for a litre of algal biofuel cannot compete with US$100 for a barrel of fossil fuel (Park *et al.*, 2011, p.35; Singh *et al.*, 2011, p.28).

Marine microbial fuel cells are devices that convert chemical energy into electrical energy by using the metabolism of microorganisms. Experimental setups involve electrodes that are immersed in marine sediment and seawater. The oxidation of organic matter in the marine sediment at the anode results in the production of electrons that travel to the cathode in seawater. The power generated is relatively low (0.05 W/m^2), but it is maintained over very long periods. Marine microbial fuel cells can thus power oceanographic instruments for long-term measurements. Moreover, the energy from fuel cells is renewable, non-polluting and efficient (Reimers *et al.*, 2001, p.195; Tender *et al.*, 2002, p.824; Yin and Zheng, 2008, p.S593).

Industrial enzymes

The search for industrial enzymes focuses on thermophilic bacteria that thrive near geothermal areas, such as volcanoes, hot springs, geysers and also in the deep sea at hydrothermal vents (Glowka, 1996, p.160). Vent bacteria often have to survive in extreme conditions: temperatures are high (species have been found at temperatures even exceeding 100°C); oxygen levels are low; the pressure is high and the surrounding waters are contaminated with compounds that are toxic for most organisms, such as hydrogen sulfide, acids and heavy metals (Deming, 1998, p.283; Eichler, 2001, p.263). Enzymes from vent bacteria, or 'extremozymes', have considerable potential for industrial processes, which involve enzymatic reactions and require high heat, pressure and extremes of pH (Leary, 2006, p.160). Such enzymes could be used for various biotechnological applications involving the amplification of deoxyribonucleic acid (DNA), waste treatment, paper processing, mining and many other fields (Glowka, 1999, p.57; van den Hove and Moreau, 2007, p.50).

The most obvious example is heat-stable DNA polymerases, which are enzymes responsible for elongating primer strands of DNA molecules. These polymerases are applied for the polymerase chain reaction – a technique for amplifying DNA involving temperatures as high as 94°C. The biotechnology company New England BioLabs Inc. manufactures several commercial DNA polymerases from deep sea species: Vent$_R$® DNA Polymerase, which derives from the archaea *Thermococcus litoralis*; Therminator™ III DNA Polymerase from *Thermococcus* sp and Deep Vent$_R$™ DNA Polymerase from *Pyrococcus* sp (New England Biolabs, 2012). All three have the advantages that they are highly thermostable and have an error-rate that is 5 to 15-times lower than other DNA polymerases from terrestrial sources (Mattila *et al.*, 1991, p.4967). The potential market for industrial uses of compounds derived from thermophilic bacteria in general lies between US\$600 million–3 billion annually (Glowka, 1999, p.58).

Mariculture

Aquaculture is the fastest growing food-producing sector in the world, with important contributions to economic and social welfare. Global aquaculture is growing at 8.8 per cent annually. In 2004, global aquaculture produced almost 60 million metric tonnes of seafood, worth over US\$70 billion (FAO, 2004, p.1). Mariculture, which is the cultivation of marine plants and animals, produced over 31 million tonnes valued at US\$31.7 billion in 2005. The top products, based on quantity produced, are kelp and molluscs. The mariculture of Atlantic salmon, however, generated the highest revenue at over US\$4.6 billion (Engle, 2009, p.545; Phillips, 2009, p.37). Most products from aquaculture have been traditionally bred, but there is increasing commercial interest in applying modern biotechnologies to enhance aquaculture.

The discussion about modern biotechnologies and mariculture started more than two decades ago. The improvement of mariculture by means of modern biotechnology has been required to address: the increasing public demand for seafood; economic inefficiencies caused by insufficient control of reproduction, development, and growth of organisms as well as high losses resulting from little resistance against environmental changes and infection by pathogens (Morse, 1986, pp.348–353).

Transgenic fish is one example. A transgenic fish is a genetically modified fish that carries a transgene – a genetic construct extracted from a source organism and integrated stably into the genome of the target organism (Chiou *et al.*, 2007, p.833). Such modified fish can be more resistant to low temperatures and pathogens and have improved growth performance (faster growth, larger adult size and increased food conversion efficiency) (Maclean, 1998, p.260; Hew and Fletcher, 2001, p.200; Melamed *et al.*, 2002, p.260; Dunham, 2004, p.172).

Concerning improved growth, one transgenic fish, originally an Atlantic salmon, *Salmo salar*, is pending to receive official approval for human consumption in the United States (Adams, 2010). The Food and Drug Administration is considering approving the request by AquaBounty Technologies to market its transgenic salmon under the brand AquAdvantage® salmon (FDA, 2010, p.10; AquaBounty Technologies, 2012). AquAdvantage salmon carry transgenes comprising an antifreeze promoter from Atlantic ocean pout, *Zoarces americanus*, and a growth hormone gene from Pacific chinook salmon, *Oncorhynchus tshawytscha*. Their expression results in a higher concentration of growth hormones, which in turn results in salmon that reach market weight twice as quickly and require 25 per cent less feed than non-modified Atlantic salmon in mariculture (Cook *et al.*, 2000, p.17).

Neutraceuticals and food-additives

The marine environment is also a rich source of neutraceuticals and food-additives. Neutraceuticals, which combines the terms 'nutrition' and 'pharmaceuticals', are specific foods or food-supplements that provide health benefits (Concise Oxford English Dictionary, 2006). Examples include Spirulina Pacifica® marketed by Cyanotech Corporation, with revenues exceeding US$8.7 million. Spirulina Pacifica is a mixture of proteins, carotenoids, B-vitamins and amino acids yielded from culture of a specially bred strain of the blue-green microalga *Spirulina platensis*. Cyanotech also markets BioAstin®, which contains the carotenoid astaxanthin that exhibits strong antioxidant activities. BioAstin is produced by culturing the microalgae *Haematococcus pluvialis*. Annual sales of BioAstin generate revenues of almost US$16 million (Cyanotech, 2012, pp.4–6).

Another prominent example is long-chain polyunsaturated fatty acids, particularly docosahexaenoic acid (DHA). DHA has many health benefits ranging from improving cognitive abilities in infants to the long-term maintenance of cognitive functions in adults and reducing the risk of dementia, depression, primary cardiac arrests, certain types of cancer, inflammation and asthma (Horrocks and Yeo, 1999, pp.212–222; Willatts and Forsyth, 2000, pp.95–99). For a long time, the primary source of DHA for human consumption has been fish oil (Kim and Mendis, 2006, p.386). However, the actual producers are microalgae, which are ingested by fish (bioaccumulation) (Liles, 1996, p.253). Industry takes advantage of this fact by applying microalgal fermentation to produce large amounts of DHA. For example, Martek cultures the microalgae *Crypthecodinium cohnii* and uses DHA for *life'sDHA*™ (DHASCO®), which is a supplement for various food products (Ward and Singh, 2005, p.3631; Martek, 2009).

Food-additives are substances added to food in order to preserve flavour or influence its appearance. Marine molecules used as food-additives are mainly polysaccharides. These are commonly applied as thickening agents, which include alginates produced by brown macroalgae as well as agars and carrageenans, both produced by various species of red macroalgae. The market is extremely broad covering coatings, adhesives, feed stocks, substrates, pharmaceuticals and biotechnological separations, and the combined annual value exceeds US$1 billion (Radmer, 1996, p.266; Weiner *et al.*, 1985, p.899; Foresight Marine Panel, 2005, p.47).

Personal care

Products in the personal care sector include skincare, hair care, cosmetics, oral care, personal hygiene and fragrances (Beattie *et al.*, 2005, p.278). Personal care products that contain natural ingredients are increasingly popular, and global sales reached almost US$8 billion in 2010, with estimated growth rates of 8–10 per cent annually (Meisel, 2009, p.6; ten Kate and Laird, 2000b, p.264). Most natural ingredients are terrestrially derived, but various products also include marine extracts (Spolaore *et al.*, 2006, p.91).

A Caribbean gorgonian species, *Pseudopterogorgia elisabethae*, is the source organism for potent anti-inflammatory and analgesic compounds called pseudopterosins. A specific compound, pseudopterosin A, is the active ingredient in various cosmetic skin-care products marketed by Estée Lauder under the brand Resilience® (Kijjoa and Sawangwong, 2004, p.81; Heckrodt and Mulzner, 2005, p.19; Estée Lauder, 2012).

The extremophile bacterium *Thermus thermophilus* is the marine source for a complex of heat-stable extremozymes. The bacterium, which was originally discovered in the Gulf of California, thrives at temperatures above 70°C and pressures from 200 to 400 atmospheres. Its extremozymes eliminate free radicals, protect cell structures from UV damage and reduce the signs of photo ageing. They are produced by industrial fermentation, at high temperatures and pressure, and used in the skin-care product Venuceane™ developed by Sederma, a subsidiary of Croda International (Leary *et al.*, 2009, p.191; Lintner *et al.*, 2009, p.463; Sederma, 2012).

Other potentially useful compounds are mycosporine-like amino acids, which absorb UV light and thus prevent DNA damage. Over 20 different compounds occur in nature, and over 400 marine species produce mycosporine-like amino acids, such as cyanobacteria, algae, invertebrates, ascidians and fish. The ability of mycosporine-like amino acids to act as effective sunscreens has attracted the interest of the commercial sector, which is studying their application as sunscreens for human skin (Karentz, 2001, pp.482–93).

Pharmaceuticals

The well-being of human society strongly depends on pharmaceuticals. Their importance is reflected in worldwide sales, which exceeded US$825 billion in 2010 (Dow Jones, 2010). For a long time, natural bioactive molecules were an important source for the research and development of pharmaceuticals and constituted the basis for 57 per cent of the 150 most prescribed pharmaceuticals (Sennett, 2001, p.523). However, the interest in developing pharmaceuticals from natural sources declined owing to high costs in time, money and expertise, as well as the low success rate associated with the development of commercial products (Ortholand and Ganesan, 2004, p.271). In addition, advances in combinatorial chemistry, that is synthesis or computer simulation of a large number of different molecules, in conjunction with improving analytical methods, such as high-throughput screening, prompted many pharmaceutical organizations to scale down or even terminate their research efforts on natural compound pharmaceuticals in the last two decades. The result was a drop to 26 per cent of natural-product-derived drugs in 2002 (Butler, 2004, p.2141). However, the rush for alternative methods to natural product research did not result in the expected development of a significant number of commercial products. On the contrary, the actual hit rate for synthetic compound libraries was lower than 0.001 per cent, with only one commercial drug eventually being commercialized (Li and Vederas, 2009, p.162; Newman and Cragg, 2007, p.461). The low success rate of combinatorial chemistry and improved biotechnological methods concerning analysis and purification of extracts have together caused a resurge in natural product research in the last six years (Molinski *et al.*, 2009, p.69; Rishton, 2008, p.43D). Marine natural products have attracted particular attention due to a hit rate in anti-cancer activity of 1 per cent, which is significantly higher than the hit rate of 0.01 per cent for terrestrial samples (de la Calle, 2009; Arico, 2006, p.19). The higher hit rate can be explained by the abundance of sessile, soft-bodied marine organisms that depend more on chemical defence, attack and signalling compounds than terrestrial organisms (Thakur *et al.*, 2005, p.472; Changyun *et al.*, 2008, p.2321–2324). In the course of evolution, such organisms evolved compounds against highly specific biological targets in competitors, and these can also affect targets in humans (Newman *et al.*, 2008, p.143).

Literature abounds on the isolation and identification of new pharmacologically-active molecules from marine organisms, which show antibiotic, anti-cancer, anti-infective, anti-inflammatory, anti-malarial, anti-viral, anti-pain, anti-Alzheimer and bone-healing activities (Proksch *et al.*, 2002, p.125; Haefner, 2003, p.537; Frenz *et al.*, 2004, p.17; Fenical, 2006, p.111). Source organisms include mainly sponges, cnidarians, tunicates and microorganisms, but also molluscs, bryozoans, echinodermata, algae and fish occurring in habitats ranging from deep-sea microhabitats to coastal areas (Carté, 1996, p.272; Minh *et al.*, 2005, p.307). To date, more than 30,000 marine natural products have been identified and each year, more than 1,000 new compounds are added to this number (Blunt *et al.*, 2010, p.164; CHEMnetBASE, 2012c).

The exploration of marine natural products has resulted in the development of nearly 50 compounds in advanced preclinical and clinical trial (Newman and Cragg, 2004, p.1217). Most of these compounds are now completely produced by chemical synthesis; the original natural compounds usually serve as 'scaffolds' that can be built upon (Hunt and Vincent, 2006, p.58; Kennedy, 2008, p.25). Once the molecule structure and active sites are identified, chemical manipulation creates a synthetic compound with improved bioactivity, stability, solubility and lower toxicity. Thus, a final product often has little similarity to the original compound extracted from nature (CBD ABS GTLE, 2008b, p.14). Once a product has successfully passed all clinical trials, it can be made available on the market.

There are several examples of marine-derived pharmaceuticals on the market. The first marine-derived pharmaceuticals on the market all originate from different nucleosides isolated from a Caribbean sponge species, *Cryptotethia crypta* (also *Tethya crypta* or *Cryptotheca crypta*) and were analysed in the early 1950s (Bergmann and Burke, 1955, p.1501). These nucleosides – spongothymidine and spongouridine – served as scaffolds for the development of various active ingredients in several drugs (Pomponi *et al.*, 2007, p.24). Zovirax® (active ingredient: acyclovir) is an anti-viral drug used against cold sores. It is marketed by GlaxoSmithKline with around US$174 million in global sales in 2011 (GSK, 2011, p.224). GlaxoSmithKline also markets several HIV therapeutics: Combivir®, Retrovir® and Trizivir®. These drugs all contain zidovudine (azidothymidine) as an active ingredient. Sales of Combivir and Trizivir in 2011 reached US$514 million and US$201 million, respectively (GSK, 2011, p.224). The active ingredients Ara-A (adenine arabinoside)[4] and Ara-C (arabinofuranosyl cytidine) are used by Pharmacia, a Pfizer subsidiary, in various products. Ara-A or vidarabine serves as an anti-viral ingredient in Vidarbin Thilo® with estimated annual sales of around US$93 million in 2007. Ara-C is an anti-cancer compound for the treatment of leukaemia in Cytosar-U®, Alexan® or Udicil®, also with annual sales around US$93 million (Fore*sight* Marine Panel, 2005, p.74).

The venoms of predatory cone snails (conotoxins) are small peptide chains with a sequence of only 20–30 amino acids. Their extraordinarily high neurotoxicity, blocking specific calcium-ion channels that play an important role in transmitting pain signals, attracted the attention for the pharmaceutical industry (Olivera, 2006, p.381). Notably, the venomous compound, ω-conotoxin MVIIA, isolated from the Indo-Pacific cone snail, *Conus magus*, and analysed in 1979, is used directly as an active ingredient (ziconotide) within the commercial analgesic Prialt® (Olivera, 2000, p.74). Prialt has been marketed since 2005 by Elan with sales of US$6.1 million in 2010 (Elan, 2011, p.30).

Another important marine-derived bioactive compound is Ecteinascidin 743 (ET-743, trabectedin), isolated from the Caribbean ascidian *Ecteinascidia turbinata* (Kijjoa and Sawangwong, 2004, p.78). ET-743 is active against cancer cells by binding to DNA and interfering with cell division, genetic transcription and the functioning of the DNA repair machinery, thus leading to cell death (Takebayashi *et al.*, 2001, p.961). ET-743 is the active ingredient in Yondelis®, marketed by PharmaMar, a subsidiary of the Zeltia Group (PharmaMar, 2012). Yondelis is currently sold for soft tissue sarcoma and ovarian cancer within Europe and an increasing number of non-European states. Annual sales provided PharmaMar with US$54 million in revenues in 2009 (PharmaMar, 2009, p.74). Any future approval of Yondelis for other types of cancer, such as breast, prostate and lung cancer could increase global sales to even exceed US$300 million (Tobin, 2009).

Finally, the marine fungus *Cephalosporium acremonium*, collected in 1948 near the Sardinian coast, is the producer of cefixime, a cephalosporin antibiotic (Bhakuni and Rawat, 2005, p.13). Cefixime is the active ingredient in Suprax®, which is also distributed by Lupin Pharmaceuticals in the US market with 2009 revenues of around US$74 million (Lupin, 2012).

Reporter genes

Reporter genes encode for reporter proteins, which can be made visible over the rest of an organism's proteins via fluorescent techniques. Reporter genes are normally fused to a particular genetic sequence. Once expression of the sequence produces reporter proteins, scientists can: a) analyse the activity and strength of promoters; b) control whether transgenes are successfully integrated into a target organism and c) examine the spatial and temporal distribution of proteins within an organism (Naylor, 1999, p.749).

One of the most commonly utilized reporter genes encodes for green fluorescent proteins (GFP) (Chalfie *et al.*, 1994, p.802). GFP was isolated, along with another reporter protein (aequorin), from the Pacific jellyfish *Aequorea victoria* in the 1960s (Shimomura *et al.*, 2005, p.223). Reporter genes encoding for GFP find application for monitoring the development of transgenic fish. Transgenes carrying a promoter, a genetic sequence encoding for a particular trait and a GFP reporter gene are injected into fish eggs. Once the fish eggs, which have the advantage that they are transparent, emit green light, the transgene has been successfully integrated into the genome. An unusual example of applying reporter genes is GloFish®, marketed by Yorktown Technologies. GloFish is a brand under which transgenic zebrafish, carrying reporter genes for GFP and red fluorescent proteins, obtained from the Indo-Pacific sea anemone, *Discosoma* sp, are sold as ornamental fish (Gong *et al.*, 2003, pp.58–63; Yorktown Technologies, 2012). Under the influence of light, these fish display green, red or orange light (obtained by crossing green and red fluorescent fish). According to an unofficial source, sales of GloFish generate approximately US$3 million annually (Charles, 2005, p.14).

Taxonomy

Products from marine biotechnologies also have non-commercial applications, such as identifying taxonomic relationships by using DNA and ribonucleic acid (RNA) (CBD COP, 1998c, para. 17). One prominent example for this purpose is barcoding. Barcoding allows the identification of species by sequencing the DNA 'barcode'. The DNA barcode is a short, standardized DNA sequence of about 650 base pairs in a well-known gene. Differences in the sequence of the DNA barcode

Table 2.1 Non-exhaustive enumeration of marine natural compounds and their (commercial) derivatives. 'Natural compound(s)' state compound name or class isolated from the biological source. Active ingredient, if available, denotes compound name or class as modified by research and development. Much data is proprietary and especially sale revenues are rarely available. AUS: Australia.

Natural compound(s)	Market product (active ingredient)	Application	Biological source	Phylum/division	Origin	Company	Sales [US$ 10⁶/yr]	Mode of production
Nereistoxin	Evisect S® (Thiocyclam); Bancol® (Bensultap); Padan® (Cartap)	Agrochemical	Lumbriconereis heteropoda	Annelid	Red Sea	Arysta LifeScience Corp, Nippon Kayaku; Takeda Chemical Industries Ltd.	7; —	Synthetic
Halogenated furanones	Pearlsafe®; Netsafe®	Antifoulant	Delisea pulchra	Red alga	Indo-Pacific	Wattyl	550,000 (all AUS paints)	Synthetic
Dichlorooctyl-isothiazolon	Sea-Nine 211™	Antifoulant	Eunicea sp	Cnidarian (soft coral)	Caribbean	Rohm and Haas	—	Synthetic
Rhamnolipid	JBR425 (rhamnolipid surfactant)	Bioremediation (among others)	Pseudomonas aeruginosa	Bacterium	ubiquitous	Jeneil Biosurfactants	—	fermentation
DNA polymerases	Vent$_R$® DNA Polymerase; Therminator™ DNA Polymerase; Deep Vent$_R$™ DNA Polymerase	DNA amplification	Thermococcus litoralis; Thermococcus sp; Pyrococcus sp	Archaea	hydrothermal vents	New England Biolabs	—	fermentation of genetically engineered E. coli
opAFP-GHc (genetic construct)	AquAdvantage® salmon	Mariculture	Salmo salar (target genome); Zoarces americanus (promoter); Oncorhynchus tshawytscha (growth hormone gene)	Fish	Atlantic; Atlantic; Pacific	AquaBounty Technologies	—	Genetic engineering, microinjection, aquaculture

Natural compound(s)	Market product (active ingredient)	Application	Biological source	Phylum/division	Origin	Company	Sales [US$ 10^6/yr]	Mode of production
Various (vitamin B 12, proteins, polysaccharides)	Spirulina Pacifica®	Neutraceutical	Spirulina platensis	Cyanobacteria	-	Cyanotech	8.7 (2011)	Aquaculture
Astaxanthin	BioAstin®	Neutraceutical	Haematococcus pluvialis	Microalga	-	Cyanotech	16 (2011)	Aquaculture
Docosahexaenoic acid	Life'sDHA™ DHASCO®	Neutraceutical	Crypthecodinium cohnii	Microalga	-	Martek	300 (2009, overall Martek)	Fermentation
Pseudopterosin A	Resilience®	Skin cream	Pseudopterogorgia elisabethae	Cnidarian	Caribbean	Estée Lauder	-	Synthetic
Superoxide dismutase	Venuceane™	Skin cream	Thermus thermophilus	Bacterium	Gulf of California	Sederma	-	Fermentation
	Zovirax® (acyclovir)	Antiviral (herpes)	Cryptotethia crypta	Sponge		GlaxoSmithKline	174 (2011)	
	Arasena-A® (Ara-A)	Antiviral (eye)	(Cryptotheca crypta, Tethya crypta)			Pharmacia	93 (2007)	
Spongothymidine, spongouridine	Combivir® (zidovudine)	Antiviral (HIV)			Caribbean	GlaxoSmithKline	514 (2011)	Synthetic
	Trizivir® (zidovudine)	Antiviral (HIV)	Eunicella cavolini	Cnidarian		GlaxoSmithKline	201 (2011)	
	Cytosar-U® (ara-c, cytarabine)	Anti-cancer (leukaemia)				Pharmacia	93 (2007)	
ω-conotoxin MVIIA	Prialt® (ziconotide)	Analgesic (painkiller)	Conus magus	Mollusc	Indo-Pacific	Elan	6.1 (2010)	Synthetic
Ecteinascidin 743	Yondelis® (trabectedin)	Anti-cancer	Ecteinascidia turbinata	Ascidian	Caribbean	EU: PharmaMar Global: Ortho Biotech	54 (2009)	Synthetic
Cefixime	Suprax®	Antibiotic	Cephalosporium acremonium	Fungus	Mediterranean	Lupin (USA)	74 (2009)	Semi-synthetic
Green fluorescent protein gene	GloFish®	Reporter protein	Aequorea victoria	Cnidarian (jellyfish)	Pacific	Yorktown Technologies	3 (2007)	Transgenic fish aquaculture
Red fluorescent protein gene			Discosoma sp	Cnidarian (anemone)	Indo-Pacific			

allow researchers to identify known species, discover new species and relate species according to the taxonomic hierarchy. To date, barcoding has analysed over 850,000 specimens from over 100,000 species. The primary benefit from barcoding is the identification of biological diversity, which in turn facilitates the control of pests and disease vectors, the monitoring of illegal trade of species, the protection of endangered species and the measurement of environmental health, etc. (Schander and Willassen, 2005, p.79; Vernooy *et al.*, 2010, p.1).

Other

Marine-derived compounds are applied in many more fields, including as biomaterials (adhesives, ceramics, nanostructures), biocatalysts (enzymes, substrates, biopolymers) and additives for paper and textile industries (CBD COP, 2000c, paras 42 and 45; Ali and Llewellyn, 2009, p.567; Kornprobst, 2010, p.508). A complete assessment of all fields is far beyond the scope of this work.

Distribution of marine species

In the search for useful biological molecules from the marine environment, many studies have focused on potential groups of source species, such as sessile species, or interesting ecosystems (Harper *et al.*, 2001, pp.8–23; UN, 2007, paras 134–36 and 169–82). However, little has been said about the actual geographical extent of single species and the probability that a source species occurs beyond the maritime zone of a single state, which relates chiefly to ocean processes and the physical properties of seawater.

The density of seawater is 854 times greater and its viscosity is 60 times greater than that of air. This involves buoyancy and a slower sinking rate of particles from the surface, which allows organisms to spend much of their lifetime in the water column and to disperse over wider areas using less energy than terrestrial organisms. The predominant type of boundaries are density boundaries created by temperature or salinity gradients. Owing to their variable nature, density boundaries are more mobile, less pronounced and thus less inhibitive to the migration of organisms than are geographical boundaries, such as mountain ranges or large rivers, for terrestrial species (Strathmann, 1990, p.197; Lasserre, 1994, p.105; Gray, 1997, p.157; Carr *et al.*, 2003, p.S91).

Ocean processes that influence the distribution of species are mainly transport forces, such as currents, gyres and confluences. These forces enable the widespread horizontal transport of material and organisms, thereby facilitating the dispersal and interconnection of even distant ecological systems within the marine environment. This is especially important for many sessile species, since the planktonic phase of their larvae and juveniles is the only opportunity to migrate and colonize remote areas (Steele, 1985, p.357; 1991a, p.181; 1991b, p.474; Norse, 1993, p.9).

Notwithstanding exceptions, marine species tend to have regional, transregional or even global distributions, the latter applying arguably for microbial species in general. Giving credit to the controversial hypothesis that 'everything is everywhere:

but the environment selects', microbial species are ubiquitous. The abundance and small size of microbes has allowed them to disperse globally via ocean currents. However, as long as environmental conditions are not favourable, these species remain dormant and proliferate only as soon conditions permit (Fenchel and Finlay, 2004, p.780; O'Malley, 2007, pp.647–51; Cermeño *et al.*, 2010, p.5; Hughes Martiny *et al.*, 2006, p.104). As a result, their biogeography does not correspond to the maritime zones of single states but rather straddles the maritime zones of multiple states (Brundtland, 1987, p.262; Freestone, 1995, p.92).[5]

This fact complicates legal regulation that aims at controlling sampling events of species in single states, because it neglects the exploitation interests of all other states that are involved in the same species' distribution.

Conclusions

The marine environment is a rich and important source of genetic sequences, proteins, carbohydrates, lipids and other biological molecules useful for marine biotechnology. First, marine species richness is very high, which entails a very high chemical diversity of biological molecules. Second, marine biological molecules are often highly potent, because they must be effective in small concentrations and function in extreme environments. This makes them useful for fields that also require high effectiveness in low dosages, such as medicine, or extreme conditions, such as industry. In addition, such molecules often have novel, very specific biological activities, which differ from those of terrestrially-derived molecules. They are therefore highly useful for biological targets that have evolved resistance to terrestrial compounds, such as agricultural pests and pathogens. Together with the large profits available from commercializing existing products, the oceans seem to be a large pool of undiscovered, potent compounds with lucrative potential.

However, the view that the oceans are a source of 'blue gold' that can be simply harvested is wrong (Greer and Harvey, 2004, p.46). First, owing to the special characteristics of seawater, wide areas are difficult to access. Second, raw samples often require substantial investments of time and money before research and development generates a marketable product. Third, although the probability of successfully developing products from marine compounds is higher than from terrestrial compounds, the process is still very risky. Many trials may never result in a marketable product. Fourth, the distribution of many marine species is not confined to single maritime zones but straddles them according to habitat suitability and migration patterns. This may prompt other states that share the species' distribution (and hence in the species' genetic value) to challenge the sovereignty of single states to unilaterally control access and claim an exclusive share of benefits.

Despite these drawbacks, scientific and commercial interest in marine species continues to rise and sampling activities in the marine environment will only intensify (Pisupati *et al.*, 2008, p.64). As a result, conflicts between providers and users of such resources may emerge, and the legal regulation of activities becomes imperative.

3   Access and benefit sharing in the marine realm

This chapter first examines the international legal framework regulating activities using marine genetic resources, second provides examples of national implementation and third, analyses case studies on the access to marine genetic resources.

Introduction

For many centuries, terrestrial organisms have served as the primary source of genetic resources for the research and development sector. With the advancement of modern scuba diving technologies in the middle of the last century, marine organisms began to attract attention from the research and development sector. As a result, the same conflicts of interest that had made the utilization of terrestrial genetic resources problematic were extended to marine genetic resources. This section provides a rough overview of the commercial potential of marine genetic resources, the stakeholders, potential conflicts and the relevant international instruments regulating the use of these resources.

Marine organisms, biological molecules and benefits from utilization

The oceans are a prosperous source of biological molecules potentially useful for human fields of application. Marine organisms have adapted to environmental influences such as pressure, heat, pH, light, salinity and toxicity, but also to inter-species interactions, such as attack, defence and signalling. As a result, they have developed an array of specialized molecules with chemical, physical and biological properties that improve their survival chances within the marine environment. Together with a high taxonomic diversity, this makes the oceans a huge reservoir of as yet unknown biological molecules with unique effects. Recent advances in marine biotechnology have opened new opportunities to access and utilize marine organisms and their molecules as the raw materials to develop new products in the fields of basic research, ecology, medicine, cosmetics, industry, nutrition, mariculture and pollution removal (Cordell, 2000, pp.463–480; Halvorson and Quezada, 2009, pp.3942–3949).

Products yield a variety of benefits, such as the increase of fundamental scientific knowledge in marine biology, but they may also deliver financial revenues produced from marketing commercial products (ten Kate and Laird, 2000b, p.26). If a product sells successfully on the market, it could have considerable economic potential (de la Calle, 2009, p.209). For example, anti-cancer pharmaceuticals derived from marine organisms are worth US$1 billion (Cole, 2005, p.1350); the Indo-Pacific cone snail, *Conus magus*, is the source of Prialt, an analgesic, producing US$6.1 million in sales in 2010 (Elan, 2011, p.30); the Caribbean sponge *Tethya crypta* yields a variety of pharmaceuticals approximating US$900 million in annual sales by 2011 (GSK, 2011, p.224), and the Caribbean tunicate *Ecteinascidia turbinata* delivers the raw material for Yondelis, an anti-cancer compound, with estimated sales exceeding US$300 million (Tobin, 2009).

Such potential benefits, whether monetary or non-monetary, mean various stakeholders become interested in the collection and utilization of marine organisms for the development of new products.

Stakeholders

Stakeholders interested in collecting and utilizing marine genetic resources can be identified on both a national and sub-national level. On the national level, the main stakeholders are the user, the provider and the intermediate state (Grajal, 1999, p.6). These states, in turn, have sub-national stakeholders under their jurisdiction, which include users, providers and intermediaries (Afreen and Abraham, 2008, pp.5–8).

Users are any entities that collect and utilize marine organisms (or parts thereof) for the development of new products. Users can be individual scientists, private companies from various biotechnological sectors (industrial, breeding, healthcare, nutrition, etc.), private or public research institutes, universities, gene-banks, cultural collections, aquaria and other entities (Swiderska, 2001, p.11; Tvedt and Young, 2007, p.12). Users invest substantial resources – time, money and expertise – into research and development and thereby hope to generate monetary and non-monetary benefits (Jeffery, 2002, p.790; de Jonge and Louwaars, 2009, p.53).

A provider is anyone who is entitled under national law to give a user the legal right to collect and utilize marine organisms found within the jurisdictional limits of the provider state. The term 'provider state' is a short form for the concept of 'country providing genetic resources' as defined under Article 2.5 of the Convention on Biological Diversity (CBD). The CBD distinguishes between a) countries supplying genetic resources collected from in-situ sources and b) countries supplying genetic resources taken from ex-situ sources, which may or may not have originated in that country. If a country supplies genetic resources that have not originated in that country, Article 15.3 clarifies that this country must have acquired the genetic resources in accordance with the CBD provisions on access and benefit sharing (ABS). Providers may include: governments and their agencies on a local, state, regional or national level; private or public scientific institutions; conservation facilities such as zoos, aquaria and gene-banks; indigenous and local communities and other entities that administer certain sea areas and control the access to resources therein (CBD WG-ABS, 2001, p.23).

The term intermediary, or middleman, refers to any entity between the initial provider and the final users and may thus function as both. The role of an intermediary can be either to broker access between a provider and a user, or to provide scientific services to users regarding collection and utilization of marine organisms. Like user entities, intermediaries may include universities, research institutes, genebanks, but they also include professional brokers (ten Kate and Laird, 2000b, p.5).

Other stakeholders that are indirectly involved in the above processes include non-governmental organizations concerned with biodiversity and culture conservation, public organizations, non-organized elements of civil society and the media (CBD WG-ABS, 2001, p.23).

The technological capacity to collect, utilize and develop products from marine organisms and the potential to provide the marine organisms is distributed unevenly between states (McGraw, 2002, p.17; Dávalos *et al.*, 2003, p.1511). The technological capacity belongs, in the main, to the private sector and research institutes in the developed world, while the bulk of global marine biodiversity can be found within the waters of developing countries. However, it is important not to divide all countries according to this scheme, because the users and providers of a marine organism may both belong to one state, making that one nation a provider state and a user state at the same time (Ruiz, 2007, p.2), which has been also recognized in paragraph 16 of the Bonn Guidelines on Access to Genetic Resources and Fair and Equitable Sharing of the Benefits Arising out of their Utilization. The economic growth of former developing countries, such as Brazil and China, has increased their ability to collect and use their own genetic resources, and some developed countries, such as Australia and New Zealand, possess high degrees of marine biodiversity and are thus potential provider states (Lochen, 2007, p.141; Tvedt and Young, 2007, p.5). However, users from developed countries frequently have to approach providers in developing countries to obtain samples of marine organisms.

Conflicts arising from collection and utilization

This interaction between users and providers from different countries leads to conflicts of interest. Users prefer easy and unrestricted access to the marine organisms harboured by the provider state, but the interest of providers lies mainly in controlling access to their resources (Garforth and Medaglia, 2006, p.2). Users do not want to enter into additional obligations; conversely, providers want to oblige users to return a share of any benefits resulting from utilization (Svarstad, 1994, p.47).

In addition, developed states are increasingly concerned with the erosion of global biodiversity. Because the bulk of global biodiversity is located in developing countries, the main responsibility for its conservation – from the perspective of developed states – falls to developing states (Biber-Klemm, 2008, p.15). However, developing states are dependent on the exploitation of their biodiversity for employment, revenues and economic development (Edwards and Abivardi, 1998, p.242).

New problems have arisen over recent decades. Economic growth in developing countries has led to increased intra-national interaction between users and providers, which could potentially produce conflicts, such as those mentioned above. A subtler problem concerns joint ventures: scientists from developed user states may choose to cooperate with scientists in provider states in order to exploit their bargaining power to create ABS-conditions that are more favourable for them. However, an exploration of these problems is beyond the scope of this work.

International instruments

With regard to the regulation of the collection and utilization of marine organisms for the development of new products, as well as concerning conflicts that may arise, two international treaties apply (Cicin-Sain *et al.*, 1996, p.195): the CBD together with its Bonn Guidelines and the Nagoya Protocol as well as the United Nations Convention on the Law of the Sea (UNCLOS).

Convention on Biological Diversity, Bonn Guidelines and Nagoya Protocol

The CBD is the key international instrument for the conservation of biological diversity, the sustainable use of its components (biological resources) and the fair and equitable sharing of benefits arising from the utilization of genetic resources (Article 1 and preamble 4). 'Biological diversity' is, in short, the variability among living organisms at the genetic, species and ecosystem level (Article 2.1). 'Biological resources' are genetic resources, organisms or parts thereof or any other component of ecosystems that have a value for humanity (2.2). The CBD understands value to include intrinsic, ecological, genetic, social, economic, scientific, educational, cultural, recreational and aesthetic values (preamble 1).

In order to maintain biodiversity, biological resources, and their value to humanity, the CBD prescribes a multitude of instruments. These include the development of national strategies (Article 6(a)), the identification of components of biodiversity (7), the creation of in-situ and ex-situ conservation measures (8, 9) and ABS (15). ABS is shorthand for the concept of 'the fair and equitable sharing of the benefits arising out of the utilization of genetic resources, including by appropriate access to genetic resources …' (1). Genetic resources are a subset of biological resources and mean 'genetic material of actual or potential value' (2.10). 'Genetic material' is further defined as 'any material of plant, animal, microbial or other origin containing functional units of heredity' (2.9). In other words, genetic material can be any living matter, from crude DNA through cell extracts and up to whole organisms. As soon as the utilization of genetic material unlocks a value, whether potential or actual, it is turned into a genetic resource.

The concept of ABS, which is further specified within several articles of the Convention (15, 16, 19), the voluntary Bonn Guidelines and the legally-binding Nagoya Protocol, aims at satisfying the interests of all stakeholders by balancing the conservation of biological diversity with the use of genetic resources (CBD WG-ABS, 2001, p.27; Birnie and Boyle, 2002, p.571). Provider states are equipped with sovereign rights over genetic resources, which give them regulatory control over access to these resources. However, this control is qualified by the obligation to facilitate access to genetic resources in order to enable users to collect and utilize genetic resources. In return, users who seek access must share the benefits they derive from any utilization with the provider state. In this way, genetic resources become a valuable asset from which benefits may flow back to the provider country that could contribute to environmental protection as well as social and economic development (Stoll, 2009, p.3). The possibility of receiving future benefits thus incentivizes states to conserve and use biological resources sustainably in order to protect the potential for new discoveries (Farrier and Tucker, 2001, p.217). This, in turn, satisfies the conservation interest of developed states.

However, two things need to be kept in mind. First, whether the CBD, as an international treaty, places legal obligations on member states only or also on individual users and providers is a contentious issue. While some authors support only the former interpretation (Moran *et al.*, 2001, p.507), it depends on the legal system of each state. Within monist states, rules of international law apply directly to individuals and state organs. International and national law form a unit and are inseparable parts of the same legal order governing member states but also their nationals. Rules of international law can be directly applied by courts. For example, Morocco is a monist state, and international treaties become an integral part of national law through publication in the Official Bulletin (Abiad, 2008, p.103). Dualist states require a translation of international law into national law in order to cover individuals and state organs. Dualistic views support the fact that international and national law are separate legal systems, each governing a different set of addressees. International law governs conduct of member states, and national law regulates activities of nationals. Courts can only apply national law and not refer to international law (Malanczuk, 2002, p.63; Nijman and Nollkaemper, 2007, pp.2–9 and 52–62).

Second, although it is not explicitly stated, the CBD applies to terrestrial *and* marine genetic resources. Evidence is provided by the Convention text itself, which defines 'biological diversity' as the 'variability among living organisms from all sources including ... *marine* and other ecosystems' (emphasis added) (2.1). Moreover, regarding the marine environment, the Convention calls for its application to be consistent with the rights and obligations of states under the law of the sea (22.2). Finally, the work of the Conference of the Parties (COP), the Subsidiary Body on Scientific, Technical and Technological Advice (SBSTTA) and other CBD bodies also focuses on the marine environment (e.g., CBD COP, 1995a, Annex I Appendix I para 14; 1995b; CBD COP, 1998a) or on marine genetic resources specifically (CBD SBSTTA, 1996a; CBD COP, 2006; CBD WG-ABS, 2009).

United Nations Convention on the Law of the Sea

As the CBD needs to be implemented consistently with the law of the sea, ABS provisions need to be compared with rules also applicable under the law of the sea (Glowka, 1998, p.45). Since its adoption in 1982, the primary source of newer law of the sea is UNCLOS, because it prevails over the four earlier conventions on the law of the sea (Article 311.1).[1] UNCLOS does not specifically refer to 'marine genetic resources' but to 'natural resources', 'living marine resources' and 'living organisms'. There are different opinions about why UNCLOS does not specifically address marine genetic resources. Some argue that the potential of marine genetic resources for commercial applications was unknown during UNCLOS negotiations, while others argue that negotiators deliberately ignored the issue of marine genetic resources (Blaustein, 2010, p.411). Nevertheless, since UNCLOS sets out the legal framework within which all activities in the oceans must be carried out (preamble 4) (UNGA, 2000) and the above terms ('natural resources', etc.) are broad enough to cover marine genetic resources (Glowka, 1996, p.168), the relevant provisions of UNCLOS also regulate the access to and utilization of marine genetic resources (CBD SBSTTA, 2003a).

The main objective of UNCLOS is to establish a legal order for the seas and oceans, which will first, facilitate international communication and second promote the peaceful uses of the seas and oceans, the equitable and efficient utilization of their resources, the conservation of their living resources and the study, protection and preservation of the marine environment (preamble 4).

To achieve these objectives, UNCLOS establishes complex and comprehensive regimes that regulate the activities of all states within horizontally and vertically divided maritime zones. Several of these regimes – namely those on the exploration and exploitation of living resources, marine scientific research and technology transfer (CBD WG-ABS, 2009, p.15) – contain provisions with a similar effect to those within the ABS regime and would therefore seem suitable for the regulation of access to and the utilization of marine genetic resources (Gorina-Ysern and Jones, 2006, p.224). The UNCLOS living resource regime equips a coastal state with sovereignty and sovereign rights over natural resources within its maritime zones, thereby providing regulatory control over access to marine genetic resources. The marine scientific research regime prescribes complex consent procedures that carefully balance the rights and obligations of both the coastal and researching state. The technology transfer regime generally obliges the researching state to share data, results and technology with the coastal state. In summary, UNCLOS compartmentalizes the oceans into maritime zones and stipulates the rights and obligations of states with divergent interests in these zones.

MARITIME ZONES

This section introduces the different maritime zones and their geographical extent under UNCLOS (Figure 3.1). This assessment is necessary because rules and obligations regulating activities using marine genetic resources differ according to the maritime zone.

UNCLOS divides maritime space vertically and horizontally into distinct maritime zones falling within or beyond national jurisdiction. Each maritime zone is subject to a specific regime of provisions with varying scales of authority for coastal states (O'Connell, 1984, p.733; Hoagland *et al.*, 2009, p.432).

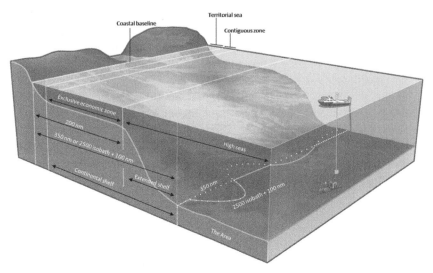

Figure 3.1 Maritime zones

Source: International Seabed Authority as reproduced in Boetius and Boetius (2011)

Vertically, the sea is divided into: the seabed, subsoil and ocean floor; the superadjacent water column and the air space. Horizontally, maritime zones are measured from baselines, which run parallel to the coast. The normal baseline is the low-water line along the coast, as marked on maritime charts and recognized by the coastal state (Article 5). Under special circumstances (coral reefs (6), straight baselines (7), river mouths (9), bays (10), low-tide elevations (13), opposite or adjacent coasts (15), archipelagic states (47)), baselines may deviate from the normal baseline.

UNCLOS describes eight main maritime zones: the internal waters, archipelagic waters, territorial sea, contiguous zone, continental shelf, exclusive economic zone (EEZ), high seas and the Area.

The internal waters lie on the landward side of the baseline, and the rights a coastal state enjoys cover the water column, the seabed and the airspace above (8.1). A similar zone is archipelagic waters, which lie within the baselines of archipelagic states (49.1).

The territorial sea extends to a maximum seaward limit of 12 nautical miles measured from the baseline (3). This covers the air space above as well as the seabed and subsoil (2.2).

The contiguous zone lies adjacent to the territorial sea and may not exceed 24 nautical miles from the baseline from which the breadth of the territorial sea is measured (33.2). Within this zone, the coastal state may only exercise control concerning customs, fiscal, immigration or sanitary laws (33.1). Resource-related rights are not covered by the contiguous zone but by the regimes on a) the EEZ or the high seas and b) the continental shelf.

The continental shelf is subject to a complex regime delineating its area. First, the continental shelf comprises the seabed and subsoil of submarine areas.

Second, UNCLOS defines the area of the continental shelf in two alternative ways, geologically or geographically (76.1). The geological continental shelf is the area beyond the territorial sea, which is the natural prolongation of land territory up to the outer margin (the margin includes the seabed and subsoil of the shelf, slope and the rise (76.3)) of the shelf. The geographical continental shelf is the seabed area beyond the territorial sea, up to a distance of 200 nautical miles, where the continental margin does not extend to 200 nautical miles from the baseline. Thus, all coastal states have a continental shelf adjacent to the territorial sea up to 200 nautical miles from the baseline, whether or not their geological continental shelf is smaller (Lagoni and Proelß, 2006, paras 64–75). This benefits oceanic islands and coastal states that have a small geological continental shelf. Where the geological continental shelf exceeds 200 nautical miles, a coastal state can extend its 'legal' continental shelf according to several geological and geographical formulas (76.4(a), 76.7). However, the outer limit of this extended continental shelf may not exceed either 350 nautical miles from the baseline or 100 nautical miles from the 2,500 meter isobath (76.5, 76.6). The existence of the continental shelf does not depend on occupation or any express proclamation (77.3). Notwithstanding its horizontal extent, the continental shelf regime does not affect the legal status of the superadjacent waters of the EEZ, the high seas or the air space above (78.1). In other words, high seas overlie the continental shelf in those areas where a) an EEZ has not been claimed and b) where the continental shelf extends 200 nautical miles.

The EEZ is the area beyond and adjacent to the territorial sea, with a seaward limit of 200 nautical miles measured from the baseline (55, 57). It thus overlaps the contiguous zone and overlies the continental shelf to a distance of 200 nautical miles. However, because the EEZ verges on the territorial seas, its actual breadth is only 188 nautical miles. The EEZ regime integrates the continental shelf regime (Lagoni and Proelß, 2006, para. 221); however, EEZ rights and obligations must be exercised in accordance with the continental shelf regime (56.3). If a coastal state has not established an EEZ, the waters superadjacent to the continental shelf remain part of the high seas.

The high seas area is defined negatively. It covers the area that is not included in the EEZ, the territorial sea or internal and archipelagic waters (86). In cases where coastal states have not declared an EEZ, the high seas verge on the territorial sea.

The Area is the seabed, ocean floor and subsoil thereof and lies beyond the limits of national jurisdiction (1.1(1)). UNCLOS does not define which areas are included within the limits of national jurisdiction. The predominant legal opinion is that the internal and archipelagic waters, the territorial sea, the EEZ (if claimed) and the continental shelf are the only areas that lie within the limits of national jurisdiction (Wolfrum, 2006, para.139). As the Area covers only the seabed and ocean floor and the subsoil, it can be best described as the area which lies adjacent to the outer limits of the (extended) continental shelf. In addition, the Area regime does not affect the legal status of the waters superadjacent or the air space above those waters (135).

Outlook

The introduction to this chapter has covered various terms, concepts and normative systems without defining, explaining or analysing them in detail. They included:

- the scope of sovereign rights and other concepts that define the legal status of marine genetic resources;
- the meaning of the terms 'genetic resources', 'utilization of genetic resources' and 'access to genetic resources';
- relevant ABS- and UNCLOS-provisions as well as their supranational and national implementation.

They will be the focus of the following sections.

Sovereign rights and the common heritage of mankind

The principles of 'sovereign rights over natural resources' and 'common heritage of mankind' are the two fundamental principles upon which the legal status of genetic resources has been developed. This section introduces both concepts and analyses how they have been incorporated into relevant instruments governing the use of genetic resources.

Background of sovereign rights over natural resources

The origin of 'sovereign rights over natural resources', as found in international law today, can be traced back to both the interests of developing states in economic development and the interests of decolonized peoples in self-determination (Schrijver, 1997, p.369). Developing states feared that resource exploitation agreements made with other states concerning their oil and mineral resources might contain conditions detrimental to their economic development. Peoples of former colonies wanted to gain political *and* economic independence from obligations that had been inherited from agreements with former colonial powers or companies. These ambitions also concerned industrialized states because resource scarcity could arise from the increased self-determination of developing states. In order to balance national interests in resource exploitation and global interests in resource availability, the United Nations (UN) has addressed this issue (Sands, 2003, p.236).

Based on the principles of the UN Charter on equal rights and self-determination (1.2, 55) and promotion of independence of territories (76(b)), the General Assembly adopted a range of resolutions which specify the rights and obligations of states concerning the use of natural resources. In 1952, the General Assembly introduced the right of under-developed states to determine freely the use of their natural resources. This right was coupled with the obligation to utilize such resources in order to realize national plans of economic development and to further the expansion of the world economy (UNGA, 1952a, preamble 1). In the same year, the General Assembly explicitly declared that the right of peoples to exploit their natural resources is inherent in their sovereignty[2] (UNGA, 1952b, preamble 3) while the phrase 'permanent sovereignty over natural resources' was employed for the first time in 1954 (UNGA, 1954, para. 1). The proceedings of the Commission on Permanent Sovereignty over Natural Resources have led to the adoption of the key resolution on 'permanent sovereignty over natural resources' (UNGA, 1962a). This, and subsequent resolutions, describes the basic rights and obligations which qualify sovereignty over natural resources (e.g., UNGA, 1970a, 1974a, 1974b, 1974c, 2009).

Although General Assembly resolutions do not have a binding effect on member states, since the General Assembly may only make recommendations according to the UN Charter (10, 11.1, 11.2, 13, 14), the principle of permanent sovereignty over natural resources reflects customary international law. This is shown by international tribunals that have ruled on the nationalization of foreign assets of natural resources,[3] a Security Council resolution concerned with coercive measures affecting the sovereignty of Latin American states over their natural resources (UN Security Council, 1973, preambles 1 and 4) and various other binding and non-binding international agreements referring to sovereign rights over natural, biological and genetic resources.[4]

Generally, sovereign rights over natural resources confer the ultimate right on states to govern independently their own natural resources within their territorial boundaries. Based on the sovereign equality of states, states must mutually respect these rights of self-determination and must not intervene in the internal affairs of other states. Specifically, within the framework of other principles and rules of international law, each state has the right to (Schrijver, 1997, p.390):

- possess, use and freely dispose of its natural resources according to its developmental priorities;
- determine freely, control and intervene with the prospecting, exploration, exploitation, development, transformation, use and marketing of its natural resources at any time;
- regulate foreign investment and the activities of foreign investors;
- determine property rights concerning natural resources, i.e., nationalize, expropriate or transfer ownership of foreign property, subject to international law requirements.

With regard to the management of marine genetic resources, this means that the resource state: a) may exclusively exploit its own marine genetic resources, and b) has the sole authority to regulate the exploitation of such resources by any other states. However, such exclusive rights are often limited by more specific soft or hard law.

Limited sovereign rights and emergence of international environmental law

During the 1960s, the international community increasingly realized that excessive resource depletion and environmental degradation caused by human exploitation might not only jeopardize economic development but also endanger the wider human environment (UNESCO, 1962; UNGA, 1962b, preamble 3–4). In response, attempts to regulate human behaviour to protect the environment began to emerge on the international scene. This marked the advent of international environmental law, which has affected and gradually changed the traditional perception of unlimited state sovereign rights (Wolfrum, 1996, p.378). What followed was the formation of an extensive body of international treaty law, case law and soft-law instruments, all of which limit sovereign rights over natural resources by novel principles arising from international environmental law (Paradell-Trius, 2000). One of the main principles includes the adoption of national environmental policies (Schrijver, 1997, pp.240–250).

Under national environmental policies, sovereign rights need to be exercised in the interest of the well-being of the people, which also depends on the quality of the environment. But states are also advised to adopt domestic legislation to preserve, restore, enrich and rationally use natural resources and to eliminate pollution. However, on its own, the obligation to adopt environmental policies is weak. The principal right to determine such policies still rests with the state, and the environmental policies of developing states should not adversely affect their developmental priorities (UNGA, 1971, paras 4(a)–(b)). This provides a broad margin of discretion for states to decide on the actual scope of their environmental policies. More substantial obligations come from a multitude of other international instruments. Non-binding rules are laid down by the World Charter for Nature, which aims at guiding human conduct which has an impact on the natural world (UNGA, 1982). Binding provisions prescribe measures on, for example, in-situ and ex-situ conservation, climate change mitigation, combating desertification, protecting the marine environment, reducing the release of persistent organic pollutants or maintaining the ozone layer.[5]

In opposition to the principle of sovereign rights over natural resources is the principle of the common heritage of mankind, which will be the focus of the following section.

Background of the common heritage of mankind principle

The common heritage of mankind principle originates from a statement submitted in 1967 by Arvid Pardo, the Maltese ambassador to the First Committee of the UN General Assembly (UNGA, 1967). In his statement, he argued that the technological capacity of a few states would soon be sufficiently advanced to access, occupy and use the deep seabed beyond national jurisdiction for the exploitation of mineral resources and strategic military purposes. The unrestricted and virtually exclusive use of the deep seabed by these few states could lead to dangerous results with potentially international consequences. First, the unrestricted exploitation of deep seabed resources would curtail the export market for land-based producers of the same resources. Second, technologically advanced states may try to secure their internationally dominant position and extend their jurisdiction to the deep seabed, thereby jeopardizing the exercise of traditional high seas freedoms by other states. Third, the exploration and exploitation of natural resources by a few states constitutes an immense detriment and inequity to technologically less advanced coastal states, which would intensify disparities in the distribution of global economic wealth. Fourth, pollution caused by the unregulated dumping of nuclear and other wastes entails serious complications for the marine living environment and endangers stocks and contaminates food supplies. Eventually, all these effects would increase world tensions, widen the arms race, endanger world peace and lead to intolerable injustices. In an attempt to provide a legal solution to these problems, Pardo proposed that the seabed, ocean floor and subsoil beyond the limits of national jurisdiction, as well as the resources found therein, are the common heritage of mankind. The main elements of this principle are:

- a prohibition of exclusive uses, national appropriation, sovereignty claims or the exercising of sovereign rights. Any activity is open to all states within the international community;
- the common heritage of mankind is administered by the international community through an international authority;
- any activity in the Area may only be undertaken for exclusively peaceful purposes;
- scientific research in the Area is freely permissible and its results are available to all;
- pollution and contamination of the Area is prevented and natural resources are protected and conserved; and
- the exploration or exploitation of the Area's resources should be carried out in the interests of and for the benefit of mankind as a whole, as well as for present and future generations.

Monetary benefits derived from commercial exploitation should be shared equitably among the international community, taking into account particularly the interests and needs of developing, geographically disadvantaged and land-locked states (Baslar, 1998, p.82).

With regard to marine genetic resources, the common heritage principle would produce two main consequences. First, the relevant area and its resources would be a common good exploitable by all. Any exploitation would benefit the international community at large, and developing countries would enjoy preferential treatment. Second, the marine genetic resources would be a common good worthy of protection. An international authority, which acts on behalf of the international community, would ensure that any activities comply with these two elements. But, as we shall see later, the common heritage is not applied to marine genetic resources (Wolfrum, 2006, para.139; Onwuekwe, 2007, p.28).

Comparison of sovereign rights and the common heritage of mankind

If the principles of common heritage and sovereign rights over natural resources are compared, we will find that both principles diverge in substance. The discrepancies occur particularly around the exploitation of resources. For example, while the international community is the title-holder and main beneficiary under the common heritage, single states determine the fate of natural resources under their sovereign rights (Francioni, 2006, pp.9–15). However, these principles were originally intended to complement, not contradict, each other. First, the common heritage over natural resources starts where sovereign rights end (Schrijver, 1988, p.100). That is, while sovereign rights apply within areas of national jurisdiction, the common heritage applies in areas beyond national jurisdiction. Second, both principles are regarded as the founding principles of the Declaration on the Establishment of a New International Economic Order, which promotes the redistribution of global wealth and power and thereby helps developing states to achieve a better position from which to realize their development plans (UNGA, 1974b, preamble 3; Schrijver, 1988, p.100).

The principles of sovereign rights over natural resources and the common heritage of mankind have been incorporated into various regimes governing the management of natural (and genetic) resources. With regard to the marine environment, UNCLOS integrates both concepts in a complementary way. However, other instruments that apply explicitly to genetic resources have used both concepts somewhat disjointedly. While initial instruments seemed to subject genetic resources to the common heritage of mankind, later instruments have served to qualify this perception.

Sovereignty, sovereign rights and the common heritage of mankind in UNCLOS

The General Assembly adopted resolutions that extended permanent sovereignty over natural resources to the marine environment. States gained the right to exercise their sovereignty over 'all their natural resources, on land within their international boundaries as well as those found in the seabed and the subsoil thereof within their national jurisdiction and in the superadjacent waters' (UNGA, 1972, para. 1; UNGA, 1973, para. 1). UNCLOS adopted the scope of these resolutions and adapted it to the various maritime zones.

The Convention established sovereignty over the bed and subsoil, water column and airspace above the internal waters, archipelagic waters and the territorial sea, as well as over the resources found therein (2). A state may rule exclusively on any matter concerning resource use, the preservation of the environment and the prevention, reduction and control of pollution (Hoagland *et al.*, 2009, p.432). Sovereignty in the territorial sea is only limited by the obligation of the coastal state to grant innocent passage (17).

Under UNCLOS, a coastal state has sovereign rights over the exploration, exploitation, conservation and management of natural resources, whether living or non-living and over other economic activities, such as the production of energy from wind and currents, in the EEZ (56.1(a)). Although these sovereign rights are far-reaching and extensive, they are not absolute because their exercise in the EEZ is restricted by several obligations. A coastal state must: a) ensure that the maintenance of living resources is not endangered by over-exploitation (61) and b) allow foreign nations access to the surplus of allowable catch (62.2). In addition, UNCLOS reaffirms the environmental principles of the Stockholm (Principle 21) and Rio Declarations (Principle 2). The UNCLOS version provides a coastal state with the sovereign right to exploit natural resources pursuant to its environmental policies and with the accompanying duty to protect and preserve the marine environment (Article 193). The Convention goes on to oblige coastal states to avoid damage by pollution to other states and their environment as well as in areas beyond national jurisdiction (194.2).

Coastal states have also sovereign rights for the purpose of exploring and exploiting their natural resources in the continental shelf region (77.1). However, there is no obligation to share any incompletely exploited living resources, as there is for the EEZ. For the continental shelf, the rights of a coastal state are exclusive in that no other state may explore or exploit the natural resources without the express consent of the coastal state (77.2).

This brief summary of sovereign rights and their qualifications illustrates that their scope is extensive insofar as they not only provide every power to regulate all activities over natural resources, but also any economic activities in these zones. The restrictions placed on coastal states are loose and weak, which leaves them with wide discretion to exercise their sovereign rights over natural resources. In fact, experts have already concluded that UNCLOS could place few restraints upon a coastal state which seeks to extend its limited sovereign rights to practically unlimited sovereignty over the living resources of the EEZ and continental shelf (Brown, 1994, p.221; Farrier and Tucker, 2001, p.223).

A complete different picture emerges for those parts of the marine environment which fall under the common heritage of mankind (the Area). UNCLOS incorporates the central principles laid down by the General Assembly (preamble 6): no state shall claim or exercise sovereignty over any part of the Area or its resources (Article 137.1); the Area and its resources can only be used, not owned (Schrijver, 1988, p96); all rights in the resources of the Area are vested in mankind as a whole, on whose behalf the International Seabed Authority shall act (137.2); the Area shall be used for exclusively peaceful purposes (141); activities shall be carried out for the benefit of mankind as a whole (140); marine scientific research shall promote international cooperation and benefit developing and shall support technologically less advanced states (143) and the marine environment of the Area shall be protected from harmful effects (145). Summarized, the main principles described in UNCLOS concerning the common heritage of mankind are almost identical to the principles already established by the General Assembly (Baslar, 1998, p.209).

Sovereign rights, common heritage and genetic resources

The first international instrument that explicitly dealt with genetic resources was the International Undertaking on Plant Genetic Resources. It was adopted by the Conference of the Food and Agriculture Organization (FAO) in response to the demand made by developing countries for international regulation of the exchange of plant genetic resources (Cooper, 1993, p.159). The problem was that the plant genetic resources of farmers were freely available, while access to improved varieties resulting from commercial breeding was restricted by intellectual property rights. To resolve this imbalance, the International Undertaking imposed an obligation on governments and institutions to allow the access to and export of samples of plant genetic resources of economic and social interest, especially for agriculture, for the purposes of plant breeding, scientific research or resource conservation (Articles 1, 2, 5). In addition, the samples were to be made available free of charge (5). These obligations were based 'on the universally accepted principle that plant genetic resources are a heritage of mankind and consequently should be available without restriction' (1). This implied that plant genetic resources are to be 'preserved, and to be freely available for use, for the benefit of present and future generations' (FAO, 1983, preamble 1).

The International Undertaking was soon heavily criticized for this. Although other critical components of the heritage principle, such as the prohibition of sovereign rights, were omitted in the International Undertaking, the use of the same terminology that applies for resources beyond national jurisdiction constituted an undermining of sovereign rights which was too strong for states to accept (Onwuekwe, 2007, p.29). The sovereign rights of states were further diluted by making plant genetic resources 'freely available' (Cooper, 1993, p.161). Industrialized states feared that the free availability of plant genetic resources would compromise their intellectual property rights, and developing states dreaded the risk of biopiracy (Lochen, 2007, p.30; de Jonge and Louwaars, 2009, p.42).

As a result, later FAO resolutions modified the International Undertaking by stating that, 'free access does not mean free of charge' (FAO, 1989, Article 5(a)) and that the rights of plant breeders should not be affected by the International Undertaking (Stoll, 2009, p.6). Other resolutions further qualified the heritage of mankind principle by explicitly subjecting it to the sovereignty of states over their plant genetic resources, and by recognizing the sovereign rights which states have over their plant genetic resources (FAO, 1991). This confusion of concepts was finally resolved by the International Treaty on Plant Genetic Resources for Food and Agriculture (ITPGRFA), which replaced the International Undertaking. The ITPGRFA fully avoids common heritage language and only refers to the sovereign rights that states have over their own plant genetic resources (preamble 14 and Article 10.1).

Another major instrument dealing with genetic resources is the CBD. Like the ITPGRFA, the CBD avoids any explicit reference to the common heritage of mankind principle and reaffirms the sovereign rights that states have over their biological resources (preamble 4). The CBD describes genetic resources as a subset of biological resources (Article 2), and therefore states also have sovereign rights over genetic resources. This view is clearly stated by 'recognizing the sovereign rights of States over their natural resources, the authority to determine access to genetic resources rests with the national governments and is subject to national legislation' (15.1).

Other authors have concluded that the common heritage of mankind principle has never applied to genetic resources within territorial limits (Mgbeoji, 2003, p.837). Instead, access to genetic resources within a state's territory was always subject to sovereign rights, and the CBD simply draws the contours of how international interactions concerning genetic resources could be shaped (Wolfrum, 1996, p.838). Therefore, later FAO resolutions to the International Undertaking, the ITPGRFA and the CBD did not create any new rights on genetic resources but only reaffirmed the rights which states have already had over any resources in their territory.

However, the common heritage of mankind principle does apply, albeit indirectly and with restrictions, to genetic resources under the 'common concern of humankind'.[6] The common concern was adopted in non-binding form within the third preamble of the CBD, which states that the conservation of biological diversity is a common concern of humankind. This formulation indicates that biological diversity has a fundamental value for mankind and that preservation is not only the internal affair of a single state but is instead in the general interest of the international community (Wolfrum, 1996, p.380). This means that a state has certain responsibilities and obligations concerning the conservation of biological diversity, which are imposed by the international community (Mahmoudi, 1999, p.221).

Thus, the common concern principle could serve as another instrument to limit absolute state sovereignty over resources for the good of the international community at large (Boyle, 1996, p.40). This also affects the use of biological and specifically genetic resources: if the common concern of humankind is the conservation of biological diversity, and the conservation of biological diversity is fundamentally linked to sustainable utilization of genetic resources under Article 1 CBD, then it is in the interests of mankind to employ a fully functional system for the sustainable utilization of genetic resources. In other words, the common concern provides the rationale and legal foundation for the ABS regime (Francioni, 2006, p.16; Pavoni, 2006, p.36).

Remarks

Originally, the principles of 'sovereign rights over natural resources' and the 'common heritage of mankind' were perceived as two distinct principles that complemented each other geographically but that could not apply to the same subject. UNCLOS has incorporated both principles coherently: while sovereign rights apply to natural resources in areas within national jurisdiction, the common heritage of mankind applies to resources outside these zones. However, international regimes governing the use of genetic resources have applied both principles incoherently. First regimes subordinated genetic resources to the common heritage, thus implicitly denying sovereign rights and all corollary rights for genetic resources found in areas within national jurisdiction. This caused an upsurge of both developing and developed states contesting this interpretation. As a result, later instruments introduced sovereign rights over genetic resources which function in parallel to the common heritage principle. The contemporary instruments have finally resolved this contradiction. They make genetic resources found in areas within national jurisdiction subject to the same sovereign rights that a state has over its natural resources.

The evolution of the legal status of genetic resources found in areas within national jurisdiction illustrates the difficulties the international community has when grappling with novel concepts.

Sovereign rights over genetic resources

Activities regulated within the ABS system, such as 'access to genetic resources', the 'utilization of genetic resources' and the sharing of benefits arising from that utilization all depend upon the concept of 'genetic resources', which is defined within the CBD. Yet the definition has been criticized for its ambiguity, failing to clearly and precisely delineate genetic resources from biological resources in general. As a result, it is unclear when the access to or utilization of genetic resources occurs, as opposed to the access to and utilization of other types of biological resources (Medaglia, 2004, p.194). Such ambiguities within the basic concepts of ABS could penetrate the whole system, eventually obscuring its exact scope and impairing its effectiveness. In order to address these issues, this section examines the scope of 'genetic resources' and related activities in contrast with 'biological resources'.

Definition of genetic resources

The CBD defines 'genetic resources' as 'genetic material of actual or potential value' (Article 2.10). Here, 'genetic material' stands for 'any material of plant, animal, microbial or other origin containing functional units of heredity' (2.9). Regardless of the scope and meaning of 'functional units of heredity', the reference to 'any material' is so broad that it covers any organic material from cells, tissues and other parts of organisms, up to whole organisms.

This interpretation significantly overlaps with the definition of 'biological resources', which includes 'genetic resources, organisms or parts thereof, populations or any other biotic components of ecosystems with actual or potential use or value for humanity' (2.2). As for the difference between 'genetic resources' and 'biological resources', because both are defined so broadly under the CBD, the concepts initially appear to be synonymous (Allem, 2000, p.338). But, equating genetic resources with all biological resources creates the problem that ABS provisions would apply to an extraordinarily large number of resources and activities (Tvedt and Young, 2007, p.64). However, the CBD definition also indicates that genetic resources are a subset of biological resources (CBD WG-ABS, 2008, para. 3). Therefore, an analysis of the main qualifying elements of 'genetic resources' is necessary in order to define its scope. The two main qualifying elements of the definition of 'genetic resources' are 'functional units of heredity' and 'of actual or potential value'.

Functional units of heredity

The concept of 'functional units of heredity' is problematic because it is insufficiently defined within the ABS-regime, and it is a scientific term, not a typical legal one.

The Nagoya Protocol to the CBD contains an indication of the scope of 'functional units of heredity' within its definition of 'derivative'. The Protocol defines a derivative as a 'naturally occurring biochemical compound resulting from the genetic expression or metabolism of biological or genetic resources, even if it does not contain functional units of heredity' (Article 2(e)). However, this definition does not reveal much about 'functional units of heredity'. It simply states that the genetic expression or metabolism may or may not result in material containing such units. It also indicates that derivatives originate through genetic expression from material containing functional units of heredity.

The scientific literature offers additional indications of the scope of this concept. The classic scientific view can be best understood when the elements of 'functional units of heredity' are analysed separately. *Heredity* is a broad concept and describes the interlinkage between the storage, transmission and expression of genetic information. Genetic information is instructions encoded and stored within *units* of nucleic acids (DNA or RNA). These units are termed 'genes' and have distinguishable beginning, protein-coding and ending sequences. The expression of the genetic information stored on these units is the *functional* part and describes the realization of these instructions which leads to the production of proteins for structural or biochemical functions (Rédei, 2008, pp.744–769, 862, 998). The underlying idea is a uni-directional flow of information from DNA over RNA to a protein, where one gene codes for one protein. In line with these considerations, scientific and legal authors refer to the gene being the basic physical and functional unit of heredity (Glowka *et al.*, 1994, p.21; Borém *et al.*, 2003, p.20; Chin *et al.*, 2006, p.905; So *et al.*, 2007, p.50; Greiber *et al.*, 2012, p.71).

However, this view can be challenged by modern microbiological knowledge and technology (Shapiro, 2009). First, laboratory experiments have transferred fragments of nucleic acids, both smaller and larger than genes, between species and have successfully triggered genetic expression. This may also include whole chromosomes and even living cells, depending on the method applied (ten Kate and Laird, 2000b, p.18).

Second, the flow of information is not only uni-directional, but it can also be bi-directional (Commoner, 2002, p.5). This means that DNA molecules are not stable structures with an unchangeable sequence. Instead, a multitude of different molecules, mainly proteins, is responsible for restructuring and changing DNA molecules, their sequences and thus their information content. Examples include DNA proofreading and repairing enzymes, DNA synthesizing enzymes that use genetic information from RNA to create DNA (reverse transcription) and various other processes that can be summarized under the heading of 'natural genetic engineering'.

Third, the mosaic view that one gene produces one protein seems to be invalid in many cases. The recently completed Human Genome Project led to the identification of 20,000 to 25,000 genes, which is not enough to account for the complexity of inherited traits in humans (Levy *et al.*, 2007, p.2114). That means that additional factors must give rise to the diversity of proteins. For example, RNA modification after transcription from DNA (alternative transcription or splicing) results in recombinations of sequences, which in turn results in production of proteins different from those encoded on DNA.

Fourth, the succinctly termed 'junk DNA' – non-coding DNA sequences that make up the majority of DNA (*c.* 97 per cent) – may contain many types of unknown genetic information essential for the proper genetic expression, replication, transmission and architecture of the genome (Encode Project Consortium, 2012).

Fifth, proteins are dynamic structures that may exchange information and modify their own structure themselves.

Sixth, a multitude of receptor molecules anticipate internal and external environmental changes and produce signals that influence all of the above-mentioned processes.

As a result, there is no single consistent interpretation of 'functional units of heredity'. While for some experts, these units are genes, others disagree (CBD WG-ABS, 2010, p.16). The most extreme alternative position sees functionality as the intricate networking (or flow of information) between many physical factors, such as DNA, RNA, proteins and other molecules, resulting in the genomic sequence and phenotype (Shapiro, 2009, p.22). This would expand 'functional units of heredity' to cover the sum of all interacting parts on the genome and all physical factors that control their expression and their direct products (Burton, 2004, p.3).

In conclusion, a final and definitive understanding of 'functional units of heredity' is not possible. Substantial scientific evidence points to genes being the central functional units of heredity. The question is how far the concept can be extended. While it is safe to assume that other parts of nucleic acids also contain genetic information that can be expressed and thus could qualify as units of heredity, extending this concept beyond nucleic acids is misleading and unhelpful. It is misleading because the majority of biological molecules that are not nucleic acids, such as proteins, do not fulfil any hereditary function, but only other biochemical functions. And, it is unhelpful because the utilization of proteins already triggers benefit sharing – there is no additional value in elaborating this concept.

It is therefore recommended that 'functional units of heredity' should only refer to nucleic acids and the specific parts of nucleic acids that encode for functions. Thus, genetic material can be defined as any biological material containing nucleic acids. This includes living cells in any appearance (seeds, eggs, sperm, whole multi-cellular organisms, micro-organisms, cell cultures, etc.) and parts of organisms, but also isolated DNA in the form of chromosomes, bacterial plasmids or parts thereof, genes and DNA fragments smaller than genes (Henne, 1998, p.39).[7] However, focusing only on 'functional units of heredity' is insufficient for improving the functionality of the ABS system and further concepts must be analysed.

Actual or potential value

The term 'value' does not help to limit the definition of genetic resources. The CBD perceives the value of biological diversity and its components to include intrinsic, ecological, genetic, social, economic, scientific, educational, cultural, recreational and aesthetic values (preamble 1). In addition, it is arguable that all genetic material could have 'potential' value, unless proven otherwise (Glowka *et al.*, 1994, p.22).

Nevertheless, 'value' does qualify the concept of genetic resources in one important way: because value is only created by human perception and utilization, genetic resources are not simply biological material containing functional units of heredity but are inherently tied to utilization (Tvedt and Young, 2007, pp.55–59).

Remarks

The analysis above advocates the view that genetic resources can be any biological material that contains nucleic acids and that is valuable. This insight contributes little to achieving the objectives of the CBD concerning especially benefit sharing. More important than the definition of genetic resources is an analysis of 'utilization of genetic resources', because it is the value-creating element and triggers benefit sharing.

Utilization of genetic resources and biological resources

The 'utilization of genetic resources' is the core concept of the ABS system that links the resource with the sharing of benefits (CBD ABS GTLE, 2008a, p.4). The Nagoya Protocol defines the 'utilization of genetic resources' as 'research and development on the genetic and/or biochemical composition of genetic resources, including through the application of biotechnology as defined in Article 2 of the [CBD]' (2(c)). 'Biotechnology' is defined as 'any technological application that uses biological systems, living organisms or derivatives thereof, to make or modify products or processes for specific use' (2(d)). In other words, genetic resources are the raw material for biotechnology (Tvedt, 2006, p.7). Finally, 'derivative' means a 'naturally occurring biochemical compound resulting from the genetic expression or metabolism of biological or genetic resources, even if it does not contain functional units of heredity' (2(e)).

These provisions clearly describe the range of utilization that triggers benefit sharing: 'utilization' does not only include compounds that qualify as genetic resources, i.e., those that contain nucleotide sequences, but it also captures compounds that do not qualify as genetic resources (Kamau *et al.*, 2010, p.256). Despite this extension, it is not clear which specific uses can trigger benefit sharing. The CBD and the Nagoya Protocol only refer generally to 'research and development' and 'biotechnology'. Although scientific literature on biotechnology abounds, additional legal guidance on specific uses must be sought.

More specific uses of 'utilization of genetic resources' can be found indicated in prior deliberations in the CBD. The CBD Group of Legal and Technical Experts on Concepts, Terms, Working Definitions and Sectoral Approaches developed a non-exhaustive catalogue consisting of the following criteria (CBD WG-ABS, 2008, paras 11–17):

- Genetic modification (the transfer of genetic traits, the genetic modification of micro-organisms, the production of recombinant cell lines and transgenic organisms, recombinant DNA techniques, the fusion of cells);
- Biosynthesis (the use of genetic material to produce organic compounds such as antibodies, vitamins, hormones, enzymes, active pharmaceutical compounds through genetically modified micro-organisms);
- Breeding and selection (the creation of new varieties through the sexual or asexual reproduction of plants, animals, micro-organisms or the domestication of these);
- Propagation and cultivation of micro-organisms, plants, animals (also for the production of other products);
- Conservation (captive breeding, seedbanks, genebanks, culture collections, aquaria, etc.);
- Characterization (the sequencing of genes, phenotyping, the evaluation of heritable characteristics, the creation of reference collections); and
- Production of compounds (the extraction or synthesis of metabolites, the synthesis of DNA segments, the copying of DNA segments through polymerase chain reaction).

Additional guidance for interpreting the 'utilization of genetic resources' can be taken from the Organisation for Economic Co-operation and Development (OECD). The OECD provides a more specific catalogue of activities that focuses only on modern biotechnologies (OECD, 2012):

- DNA and RNA (genomics, pharmacogenomics, gene probes, genetic engineering, DNA/RNA sequencing/synthesis/amplification, gene expression profiling and the use of antisense technology);
- Proteins and other molecules (the sequencing, synthesis and engineering of proteins and peptides, including large molecule hormones; improved delivery methods for large molecule drugs; proteomics, protein isolation and purification, signalling and the identification of cell receptors);
- Cell and tissue culture and engineering (cell and tissue culture; tissue engineering, including tissue scaffolds and biomedical engineering; cellular fusion; vaccine and immune stimulants and embryo manipulation);
- Process biotechnology techniques (fermentation using bioreactors, bioprocessing, bioleaching, biopulping, biobleaching, biodesulphurization, bioremediation, biofiltration and phytoremediation);
- Gene and RNA vectors (gene therapy, viral vectors);
- Bioinformatics (the construction of databases on genomes, protein sequences; modelling complex biological processes, including systems biology); and
- Nanobiotechnology (applying the tools and processes of nano-/microfabrication to build devices for studying biosystems and applications in drug delivery and diagnostics).

Summarizing these lists, the 'utilization of genetic resources' would cover three types of activities (Glowka *et al.*, 1994, p.17; CBD COP, 1995c, para. 7). First, activities use the reproductive ability of biological systems in order to express certain traits encoded within genetic information, which can then be further used as products. This includes any type of cultivation, culture, multiplication, propagation and replication of plants, animals, micro-organisms and cells (Glowka, 1997a, p.251; Henne, 1998, pp.39–41, 67–68, 148; Medaglia and Silva, 2007, p.26). Examples could be raising mass cell cultures for the production of medical compounds, the raising of crops and livestock for food, the breeding of plants and animals for developing novel traits or captive breeding in zoos and aquaria for conservation. It is important to note that 'utilization' covers the reproduction process and not necessarily the use of the product. Many products from reproduction are used as 'biological resources', which does not trigger benefit sharing.

Second, activities target the smallest parts which make up organisms: biological molecules. This includes a vast set of techniques that actively target cells, their DNA and RNA, proteins and other bioactive molecules in order to make or modify products and processes (Guilford-Blake and Strickland, 2008, p.18). Examples are the isolation of molecules and analysis of their structure, the analysis of molecules for their (bio-)activity and function, the modification of molecules for deleting or improving certain traits and the transfer of molecules between species. Products are manifold and include the generation of knowledge about biological compounds and their function, the identification of species, the production of molecules for specific application in various fields and the development of improved species (Bains, 2004; Guilford-Blake and Strickland, 2008, pp.23–69).

Third, 'utilization' involves digitally managing biological information (bioinformatics). This includes storing and representing biological information or modelling biological processes.

Utilization of genetic resources vs. 'traditional' uses

Distinguishing between the uses of genetic resources and other uses of biological material (traditional uses) has long been a contentious issue, and there is no clear formula to delineate both types. Thus, stakeholders may disagree about whether or not a specific use of a resource constitutes the 'utilization of genetic resources'. In turn, this could cause uncertainties for benefit sharing.

First indications can be derived from interpreting the definition of 'utilization of genetic resources'. As described above, utilization involves the reproduction of biological systems, using biological molecules for specific purposes (genes or 'functional units of heredity', proteins and bioactive molecules) (Dross and Wolff, 2005, p.54) and the information management of biological molecules. Traditional uses could be all activities that do not fall under this interpretation. The legal literature provides additional guidance: traditional uses include the direct consumption of plants, animals and parts thereof for nutrient uptake, such as meat, grains, fruits, vegetables and herbs, the felling of trees and their use as firewood or timber, using organisms as ornamentals, such as flowers or aquarium fish, the holding of a domestic pet or working animal and countless other uses (Glowka, 1997a, p.251; Henne, 1998, p.41; Young, 2004, p.48).

The above considerations support the need to identify additional criteria to distinguish the utilization of genetic resources from traditional uses:

- The key commodity of genetic resources within the ABS system is less the physical specimen but rather its biological molecules or the information they contain (sequences, structures) (Schroeder, 2000, p.55). For example, a salmon specimen with a desirable trait encoded within its genetic sequence can be used for breeding. The key commodity for traditional uses is only the physical organic material of body parts, regardless of any specific molecules (Tvedt and Young, 2007, p.55), such as an individual elephant used as a working animal.
- Genetic resources, although tangible in nature, can be transformed into intangible or informational products, without losing utility, e.g., in virtual libraries and databases containing DNA sequences. This is not possible with the majority of other biological resources (Soplín and Muller, 2009, p.1).
- As the utilization of genetic resources targets mainly the biological molecules, only small quantities of material (samples of tissue, water and soil or few organisms) are necessary to unlock their value (Korn *et al.*, 2003, p.43). On the contrary, biological resources for traditional uses are often harvested and sold in bulk in order to meet consumption needs.
- The utility and value of genetic resources is often independent from organism numbers but primarily linked to the quality of the biological molecules, which can be derived from any individual within, e.g., a species. However, the utility and value of most other biological resources is directly proportional to the physical quantity of the resource extracted (Kloppenburg, 1988, p.187).
- Theoretically, the utility of a genetic resource does not diminish once extracted, since the resource is often propagated or cultured. The utility of other biological resources depletes with their consumption, e.g., during digestion of food or death of a working animal.
- The utilization of genetic resources rarely produces commercial products (Kuhlmann, 1997, p.541), whereas the extraction of other biological resources always yields products.
- Products derived from the utilization of genetic resources regularly require a lot of processing with large investments in highly technical equipment, expertise, time and money. Other biological resources extracted as bulk for consumption require relatively little, if any, processing to make the resources economically valuable (Cicin-Sain *et al.*, 1996, p.204).
- The utilization of genetic resources is usually not limited by maximum levels available for extraction, because extraction often requires only sample quantities, the relevant organisms reproduce in high frequency (e.g., micro-organisms) or final products which can be synthesized. Although traditional biological resources are also renewable, they can constitute a finite resource,

if the techniques and equipment used for extraction yield amounts of the resource larger than natural reproduction can sustain (Korn *et al.*, 2003, p.43).

• In a similar vein, the extraction of genetic resources rarely affects natural populations, whereas the excessive extraction of other biological resources can quickly lead to overexploitation (Wolfrum and Matz, 2000, p.456; Hunt and Vincent, 2006, p.60).

Remarks

The utilization of genetic resources remains a legally challenging concept. The Nagoya Protocol has helped to build an intuitive understanding of the concept, but it fails to separate the countless number of uses of biological material clearly into those that trigger benefit sharing and those that do not – although this is an outcome highly unlikely to be achieved given the complexity of the issue. Additional guidance can be sought from catalogues of criteria and specific uses as provided by the Group of Technical and Legal Experts or the OECD, but there is no legal obligation to apply these. Another option that will be mentioned only marginally is extending the scope of ABS to cover all uses of biological resources as proposed by the African Model Legislation for the Protection of the Rights of Local Communities, Farmers and Breeders, and for the Regulation of Access to Biological Resources. While such an approach provides the opportunity for source states to outline their ABS legislation as they deem fit, it has the main disadvantages that it counteracts an internationally harmonized ABS system and creates a fragmented approach to managing genetic resources.

The definition of the 'utilization of genetic resources' within the Nagoya Protocol should be criticized for excluding traditional taxonomic identification from the uses that trigger benefit sharing (CBD SBSTTA, 2012, p.4). Traditional taxonomy relies on the morphological characteristics of specimens to identify the species. Because utilization covers only research on the 'genetic and/ or biochemical composition' of genetic resources, the benefits of traditional taxonomy do not need to be shared with the provider state. This is a great loss of a benefit which could have otherwise greatly contributed to conservation and sustainable use in provider states by identifying, e.g., rare or endangered species (CBD COP, 1999, para.77). There are three possible options for solving this issue: scientists that engage in traditional taxonomy could share their results with the provider state voluntarily (as proposed by para. 11(1) of the Bonn Guidelines), taxonomy is defined as 'special' utilization of genetic resources, or taxonomy is understood as indirect utilization of genetic resources, because the morphology of species ultimately relies on the species-specific genetic composition. The last point, however, is highly problematic. It could lead to the conclusion that every use of biological resources qualifies as a utilization of genetic resources and thus triggers benefit sharing. Provider states would then have the freedom to demand benefit sharing for every use of biological resources, which – if formulated restrictively – would enable them to limit access to their resources and curb trade for foreign companies.

Access to genetic resources

The difficulty in distinguishing the utilization of genetic resources from traditional uses also affects our understanding of 'access to genetic resources'. If 'access' is only understood to be obtaining and taking physical biological material, it is not clear if taking a particular organism constitutes access to a genetic or to a biological resource.

A utilization-based approach would be helpful in clarifying 'access to genetic resources' (Medaglia, 2004, p.197; Medaglia and Silva, 2007, p.26). Under this approach, 'access to genetic resources' would not only include the actual taking of material but would also depend on the intended or declared use of the material in question, e.g., using the material for propagation (Henne, 1998, p.148). This means that the party seeking access to biological material must provide information on the intended use. If it describes uses that qualify as a utilization of genetic resources, then access to a genetic resource is established and further negotiations can follow as intended under the ABS system.

Remarks

The concepts of genetic resources and associated activities have undergone much debate. The result is a definition that extends benefit sharing to the utilization of material that does not actually qualify as a genetic resource. Nevertheless, this extension is justified on several grounds. First, research and development on the biochemical composition of genetic resources require very similar tools and techniques to research and development that involve functional units of heredity. Second, the pharmaceutical industry relies primarily on the extraction and synthesis of biochemical compounds for the discovery and development of drugs (CBD ABS GTLE, 2008b, p.10). As it has been a major aim of the ABS system to subject this sector to benefit sharing, extending the definition to material outside the scope of 'genetic resources' is imperative.

Despite this, there are still several issues which need to be resolved. First, an official, non-exhaustive catalogue of criteria and specific uses that qualify 'utilization' would greatly contribute to distinguishing the utilization of genetic resources from other uses. Second, although traditional taxonomy is excluded from 'utilization', its relevance to ABS must be acknowledged and benefit sharing encouraged.

International instruments for activities on marine genetic resources

At present, there are two international conventions that regulate activities involving marine genetic resources. The first, UNCLOS, aims at regulating all activities in ocean space. It does not specifically refer to 'marine genetic resources', but the relevant provisions within UNCLOS can also be applied to activities involving marine genetic resources, because marine genetic resources qualify as natural living resources (Glowka *et al.*, 1994, p.47). Relevant provisions regulate the exploration and exploitation of natural resources, marine scientific research and the transfer of technology (Cicin-Sain *et al.*, 1996, p.204; CBD SBSTTA, 2003a, para. 21; CBD WG-ABS, 2009, p.15). The second regime, the ABS regime under the CBD, aims at the sharing of benefits that are derived from utilization of marine genetic resources taken from the maritime zones that belong to coastal states.

Both conventions not only complement and reinforce each other, but may also conflict in certain areas. It is therefore the aim of this section to compare the provisions of both conventions which concern activities using marine genetic resources.

A full comparison of all elements of UNCLOS and the ABS regime would quickly exceed the space available for this work. Therefore, the analysis will focus only on core concepts; a detailed assessment can be found at the end of this section in Table 3.1.

Access to marine genetic resources

The access to marine genetic resources is regulated by different legal regimes in each convention. In UNCLOS, access to such resources falls under the exploration and exploitation of natural resources or marine scientific research. The CBD treats access to marine genetic resources more generally in its ABS regime, as laid down within the Convention text, the Bonn Guidelines and the Nagoya Protocol (Greiber *et al.*, 2012). Because the rights and obligations outlined under each regime may differ according to the relevant maritime zone, the following analysis is divided into: a) general provisions that apply either independently of or to multiple maritime zones and b) specific provisions that apply to particular maritime zones.

General provisions under UNCLOS

Provisions on the exploration and exploitation of natural resources are regulated in the specific parts that also establish the maritime zones and are therefore examined separately below. However, the rights of resource exploration and exploitation do not apply if an activity on marine genetic resources qualifies as marine scientific research.

BASIC RESEARCH VS. ECONOMIC RESEARCH

Neither UNCLOS nor the preceding conventions on the law of the sea defined 'marine scientific research'. In the absence of a legal definition, it has been proposed that 'marine scientific research' is any study and experimental work designed to increase human knowledge of the marine environment. The disciplines covered by marine scientific research are marine chemistry, physics, geology, geophysics, meteorology, hydrography and biology. The latter includes broadly the study of plant, animal and microbial organisms in the oceans and therefore includes research activities on marine genetic resources (Wegelein, 2005, pp.11–16).

Research activities on marine genetic resources may have non-economic as well as economic intentions. Yet it is not clear to which activities 'marine scientific research' applies, because UNCLOS does not give an explicit distinction in this regard. To determine the scope of 'marine scientific research', it is helpful to distinguish those research activities that belong to basic research and those that are applied research.

Basic research, also termed pure or fundamental research, includes all experimental or theoretical research activities that aim simply to increase scientific knowledge without aiming at a particular application in any other field. Results from basic research provide new insights into general science and are often quickly published and freely disseminated to interested parties (OECD, 2002, paras 64, 240–244).

Applied research also aims to produce new knowledge, but with results which can be applied in other sectors. Applied research often has practical aims and serves a specific non-economic or economic purpose. If applied research activities have an economic purpose, the progress and results are commonly concealed until intellectual protection is guaranteed or a marketable product has been developed (OECD, 2002, paras 64, 246–248).[8]

In UNCLOS, marine scientific research involves the collection of information, data or samples, and is characterized by transparency, the availability of knowledge and data and the dissemination and publication of research results (Article 244) (Glowka, 1996, p.172). Thus, the marine scientific research regime primarily covers basic research but also applied research, as long as it is not conducted for any economic purposes (Soons, 1982, p.6; CBD SBSTTA, 2003a, paras 39 and 47; Wegelein, 2005, p.77; UN, 2007, para. 203; Kirchner, 2010, p.120). Applied research activities on natural resources that are conducted for economic purposes are not considered marine scientific research and instead fall under provisions concerning the exploration and exploitation of natural resources (Soons, 1982, p.125; Glowka et al., 1994, p.47), although this view is not shared by everybody (Cataldi, 2006, p.103). Exploration with regard to research activities has been described as data-collecting activities concerning natural resources conducted specifically for exploitation, i.e., economic utilization of those natural resources (Soons, 1982, p.125). Most activities on marine genetic resources, such as bioprospecting,[9] fall under this description.

MARINE SCIENTIFIC RESEARCH UNDER THE LAW OF THE SEA

With regard to marine scientific research, UNCLOS provides various general provisions and principles that apply to all maritime zones. All states have the right to conduct marine scientific research (Article 238), and both the undertaking and development of such research are to be promoted and facilitated (239). Marine scientific research is to be conducted for exclusively peaceful purposes, with appropriate scientific methods and means which comply with UNCLOS. It must also comply with regulations for the protection and preservation of the marine environment, and must not interfere with other legitimate uses of the sea without justification (240).

Further, marine scientific research activities may not constitute the legal basis of any claim to any part of the marine environment or its resources (241). The scope of this last provision has been interpreted to be potentially wide: it not only forbids claims to marine space by other states but also forbids any claims to the exclusive exploration and exploitation of natural resources, data, samples and results arising from marine scientific research (Gorina-Ysern, 1998, p.345).

UNCLOS aims at promoting international cooperation in marine scientific research (242) by obliging states to cooperate, through bilateral and multilateral agreements, to create favourable conditions for marine scientific research (243).

General access provisions under the ABS regime

The access provisions under the CBD apply to various maritime zones, namely, those that are within the limits of the national jurisdiction of contracting parties (4(a)). The CBD does not provide a more specific indication as to whether this wording covers only the internal waters and territorial sea or also includes the EEZ and the continental shelf. The vast majority of legal authors describe the continental shelf and the EEZ, despite its special legal status ranging between territorial sea and high seas (Orrego Vicuña, 1989, p.21), as areas within national jurisdiction (Brown, 1984, p.I.25; Office of the Special Representative of the Secretary-General for the Law of the Sea, 1985, pp.43–44; Glowka *et al.*, 1994, p.28; Joyner, 1995, p.646; de Fontaubert *et al.*, 1996, p.2; Wolfrum and Matz, 2000, p.470; Wolff, 2004, p.185; UN, 2007, para. 192; Proelß, 2009, p.61). This conclusion has been explained mainly by the strong jurisdictional powers of coastal states, which can be extended to the EEZ and the continental shelf under Articles 56 and 77 UNCLOS. It is therefore concluded here that the ABS provisions explained below should apply within the EEZ, if established, up to the outer limit of 200 nm and up to the (declared) outer limit of the continental shelf.

CONVENTION TEXT

Within areas of national jurisdiction, the CBD grants states sovereign rights over their genetic resources and, under national legislation, the authority to control access to such resources (preamble 4 and Articles 3 and 15.1).

Access to genetic resources requires the prior informed consent of the contracting party providing such resources, unless otherwise determined by that party (Article 15.5). The purpose of prior informed consent is so that the provider party can obtain all relevant information from the party seeking access in order to make a meaningful decision about whether or not to grant access (Lochen, 2007, p.124).

Where access to genetic resources is granted, it must be on mutually agreed terms (15.4). This provision means that the parties providing genetic resources and those seeking access must enter into negotiations to come to an agreement on the particularities of how access, subsequent utilization and the sharing of benefits are to be conducted (Glowka *et al.*, 1994, p.80). The purpose of mutually agreed terms is to balance the bargaining power between both parties through the obligation to agree to the terms (Chishakwe, 2009, p.33).

However, the party providing genetic resources is not free to demand overly restrictive conditions for obtaining prior informed consent or reaching mutually agreed terms. Each party must strive to create conditions which facilitate access to genetic resources and cannot impose restrictions that run counter to the objectives of the Convention (15.2).

In addition, the knowledge, innovations and practices of indigenous communities (traditional knowledge) play an important role for the ABS regime (8(j)). Although traditional knowledge does exist for applications of marine genetic resources (Demunshi and Chugh, 2010, pp.3025–3027 and 3031), it is much less relevant in comparison to the terrestrial context and will therefore only be treated marginally within this work (Carté, 1996, p.271; Tangley, 1996, p.245; McLaughlin, 2003, p.298).

The CBD provisions on access have been criticized for their vagueness and the ambiguity of even basic concepts which has hampered national implementation (Jeffery, 2002, p.778). In particular, the scope and details of prior informed consent and mutually agreed terms as well as their relationship are unclear. For example, it has been interpreted differently whether prior informed consent means: a) consent to access genetic resources (as a parallel requirement to the conclusion of an agreement based on mutually agreed terms) (Glowka *et al.*, 1994, p.81) or b) consent to enter into negotiations to conclude an agreement based on mutually agreed terms (Henne, 1998, p.151).

In order to develop a common understanding of these basic concepts, the CBD COP decided to establish an *ad hoc* open-ended working group (Working Group on ABS) with the mandate to create guidelines that support the implementation of legislative, administrative or policy measures on ABS (CBD COP, 1998b, para. 4; CBD COP, 2000b, para. 11). These guidelines were subsequently adopted as the Bonn Guidelines (CBD COP, 2002, para. 3).

BONN GUIDELINES

The Bonn Guidelines aim to assist governments and other stakeholders to establish legislative, administrative and policy measures on ABS (Paragraph 1). The Guidelines are divided into: a) general provisions; b) roles and responsibilities; c) stakeholder participation;[10] d) the ABS process; e) other provisions; f) elements of material transfer agreements and g) benefits. This section provides only a general overview over the main provisions, with particular emphasis on the access process.

The general provisions clarify that the Guidelines are purely voluntary and they do not alter the rights and obligations of parties under the CBD, nor do they affect sovereign rights over natural resources, nor do they affect the rights and obligations outlined in any mutually agreed terms under which genetic resources have been obtained (2, 4, 6–7). In addition, the Guidelines are intended to be simple and practical (7). Their objectives are, among others, to contribute to the conservation and sustainable use of biological diversity, to provide a transparent framework facilitating ABS, to guide the creation of ABS regimes, to assist negotiations of ABS arrangements between providers and users and to facilitate taxonomic research (11).

To achieve these objectives, the Guidelines distinguish between the roles and responsibilities that must be fulfilled by countries of origin, users, providers and user states (16). Countries of origin should ensure that their ABS measures comply with Article 15 of the CBD. Furthermore, access applications should be made available through the clearing-house mechanism. Countries of origin also need to fulfil their roles and responsibilities in a clear and transparent manner and to ensure that all stakeholders have considered the environmental consequences of their access activities. Particular emphasis is put on indigenous communities and associated traditional knowledge: commercialization should not prevent the traditional use of genetic resources, decisions must be made available to relevant communities and the ability of a community to represent their interests should be enhanced.

Users should seek prior informed consent and comply with various prescriptions on the use of genetic resources. Genetic resources may only be used for purposes consistent with the terms under which they were acquired. If this is not the case, users must obtain new prior informed consent and mutually agreed terms. Evidence that prior informed consent and mutually agreed terms have been obtained must be maintained. In addition, users should involve the provider state in utilization, share benefits fairly and equitably and pass on terms and conditions when transferring the material to third parties. Finally, users should respect local customs and respond to requests for information from communities. Providers should only supply resources or knowledge if they are entitled to do so and may not impose arbitrary restrictions on access.

User states need to adopt measures which support users to comply with the prior informed consent and mutually agreed terms on which access was granted by the provider state. Such measures include information on obligations concerning access, the disclosure of the country of origin in applications for intellectual property rights, the prevention of the use of genetic resources without consent, cooperation with parties to address alleged infringements and voluntary certification schemes.

Furthermore, it is the responsibility of parties to the Guidelines to designate national focal points and to establish competent national authorities. While the focal point is responsible for informing applicants on ABS procedures, the primary responsibility of the competent authority is to grant prior informed consent for access (13, 14, 27(a), 28). Other responsibilities of competent authorities are to provide advice on a) negotiation processes, requirements for prior informed consent and mutually agreed terms; b) monitoring, implementation and processing of ABS agreements and c) mechanisms for effective participation of stakeholders.

With regard to the prior informed consent system, the Guidelines contain various provisions on how to establish such a system. First, they outline the basics, which include (26) (Jeffery, 2002, p.797):

- legal certainty and clarity;
- access should be facilitated at minimum cost;
- restrictions on access should be transparent, based on legal grounds and not run counter the CBD's objectives; and

- consent of the competent national authorities and relevant stakeholders, as appropriate, should be obtained.

Second, the prior informed consent system should include certain elements, namely (27):

- a competent authority granting or providing evidence of prior informed consent;
- timing and deadlines, which means not only applying for prior informed consent adequately in advance but also deciding on such applications within a reasonable period of time (33);
- specific uses for which access to genetic resources are sought (34);
- mechanisms to consult relevant stakeholders;
- process including documentation in written form and transparency (38–40); and
- procedures for obtaining prior informed consent.

Procedures for obtaining prior informed consent mainly involve obtaining the information that accompanies applications. This includes (36):

- legal entity and affiliation of the applicant;
- type and quantity of the genetic resource to which access is sought;
- starting date and duration;
- geographical area of collection;
- accurate information on the uses (e.g., taxonomy, collection, research or commercialization);
- location where research and development takes place;
- purpose of collection and expected results;
- possible benefits generated from derivatives and utilization of the genetic resource;
- indication of benefit-sharing agreements;
- budget; and
- treatment of confidential information.

With regard to mutually agreed terms, the Guidelines define the basic requirements, which are similar to the provisions on prior informed consent. These include legal certainty and minimization of transaction costs by, for example, developing standardized material transfer and benefit-sharing agreements (42). Appendix I to the Bonn Guidelines contains a list of suggested elements for material transfer agreements. These involve introductory provisions, ABS provisions and legal provisions. Guiding parameters to such contractual agreements could be:

- regulating the use in order to take ethical concerns into account;
- ensuring continued customary use of genetic resources and related knowledge;

- intellectual property rights including joint research, an obligation to implement rights on inventions and the provision of licences by common consent; and
- joint intellectual property rights, depending on the degree of contribution.

The Guidelines further provide a list of typically mutually agreed terms, which include: a) the identification of capacity-building in various areas and b) whether knowledge, innovations and practices have been respected, preserved and maintained.

Finally, the Guidelines contain provisions on incentives and accountability. Incentives promote the implementation of the Guidelines and could cover valuation methods or the creation and use of markets (51). To promote accountability, parties may establish requirements such as reporting or the disclosure of information (53).

Significant effort was put into the development of the Guidelines and some major achievements have been reached. These include: a) frameworks for establishing national focal points and competent authorities; b) responsibilities for both the provider and user side and c) standardized elements for prior informed consent, mutually agreed terms and material transfer agreements.

However, the latter point was also most criticized aspect of the Guidelines. Dealing with key concepts and the introduction of novel concepts such as material transfer agreements and benefit-sharing arrangements in isolation has resulted in a fragmented picture (Lochen, 2007, pp.164–167). The consequence of this is an unclear relationship between these concepts, which also impedes the implementation of a coherent ABS system. In the legal literature it is assumed that an access seeker must first submit an application for prior informed consent. If the content of the application fulfils the requirements set by the provider state, the next step is to negotiate a material transfer agreement that executes and embodies terms mutually agreed. Only after the successful negotiation of a material transfer agreement is the provider state able to conclude the consent procedure and the user able to access a particular genetic resource (Chambers, 2003, pp.312–315; Tully, 2003, p.91; Pisupati, 2007, pp.43–49; Greiber *et al.*, 2012, p.98).

Other criticisms pertain to the non-binding nature and complexity of the Guidelines, the lack of clear definitions of genetic resources and derivatives, as well as the inadequate handling of compliance measures and transfer of technology.

A result of these shortcomings is that the Guidelines can only be another step on an evolutionary process (CBD COP, 2002, para. 6; Stoll, 2004, p.86). The Plan of Implementation of the World Summit on Sustainable Development proposed the negotiation of an international regime to the CBD, which would promote the fair and equitable sharing of benefits arising from the utilization of genetic resources (UN, 2002, para. 42(o)). In 2004, the CBD COP mandated the Working Group on ABS to elaborate and negotiate an international regime on access to genetic resources and benefit sharing, with the aim of adopting an instrument to effectively implement CBD ABS provisions (CBD COP, 2004, para. 1). In

2010, the final version of this international regime was adopted by the COP as the Nagoya Protocol (CBD COP, 2010b, para. 1).

NAGOYA PROTOCOL

The Nagoya Protocol on Access to Genetic Resources and the Fair and Equitable Sharing of Benefits Arising from their Utilization to the Convention on Biological Diversity is structured into 27 preambular paragraphs, 36 articles and one annex. The objective of the Protocol repeats verbatim the third objective of the CBD and adds that ABS shall contribute 'to the conservation of biological diversity and the sustainable use of its components' (Article 1). Thereby, the Protocol explicitly links ABS to the other two objectives of the CBD.

The access provisions under the Protocol reiterate, together with a reaffirmation of sovereign rights over natural resources, that access to genetic resources for their utilization is subject to the prior informed consent of the providing party, which is either the country of origin or a party that acquired the genetic resources in accordance with the Convention (6.1). Given experiences with over-bureaucratic and intransparent access procedures, the Protocol is very specific about the procedure for facilitating access. For this purpose, provider states shall provide (6.3):

- 'legal certainty, clarity and transparency' of their domestic ABS legislation;
- 'fair and non-arbitrary rules and procedures' for access to genetic resources;
- 'information on how to apply for prior informed consent';
- clear, cost-effective and timely decision-making;
- recognition of a permit or its equivalent as evidence of PIC;
- criteria and procedures for involvement of indigenous and local communities; and
- clear rules and procedures for requiring and establishing MAT (dispute settlement, terms on benefit sharing, intellectual property rights, third party use, change of intent) (6.3(g)(i)–iv)).

The parties on the provider side that must be involved in giving consent and agreeing on mutual terms include the provider state itself (6.1) and – according to domestic legislation – any indigenous and local communities that hold genetic resources (6.2) and/or associated traditional knowledge (7). The responsibility for advising on prior informed consent and mutually agreed terms lies with national focal points and competent national authorities (13.1 and 13.2). The latter are also responsible for granting access (13.2). One single entity may be designated to fulfil the functions of both focal point and competent national authority (13.3).

Communities that hold genetic resources and traditional knowledge associated with genetic resources are taken into consideration extensively in various provisions of the Protocol. First, where communities have the domestic right to grant access to genetic resources or hold traditional knowledge, parties must adopt measures which ensure that PIC and involvement for access is obtained from such communities (6.2 and 7). In addition, parties specify criteria and processes for obtaining prior informed consent and involving communities (6.3(f)). Second, any benefits derived from the utilization of genetic resources or traditional knowledge held by communities must be shared in a fair and equitable way with such communities (5.2 and 5.5). Third, parties, with the effective participation of communities, should establish mechanisms to inform the users of traditional knowledge about their obligations (12.2). Such obligations can be laid down in community protocols, minimum requirements for MAT and model contractual clauses as developed by communities with the support of the party (12.1 and 12.3). Fourth, in order to increase the awareness of genetic resources and traditional knowledge held by communities, parties should organize meetings of communities, establish a help desk for communities and involve communities in the implementation of the Protocol (21(b)–(c) and (h)). A first roadmap to increasing awareness on national, regional and subregional level can be found in the Awareness-Raising Strategy for the Nagoya Protocol (CBD ICNP, 2012a). Fifth, in order to enable the effective participation of communities in implementation of the Protocol, the capacities of communities also need to be improved. In this regard, the Protocol emphasizes the need to increase the capacities of women (22.3 and 22.5(j)), owing to their vital role in ABS processes, policymaking and the implementation of biodiversity conservation (preamble 11).

Although the Protocol reaffirms the sovereign rights of parties over their genetic resources, its provisions on transboundary cooperation, in cases where the same genetic resources or traditional knowledge straddle national boundaries, constitute a type of derogation, albeit weak, of absolute state sovereignty. In such cases, parties should 'endeavour to cooperate' with a view to implement the objectives of the Protocol (Article 11). The prospective 'global multilateral benefit-sharing mechanism' pursues a similar aim for genetic resources and traditional knowledge that occur in transboundary situations or for which it is not possible to grant or obtain prior informed consent (10). Such a mechanism would direct the benefits derived from the utilization of genetic resources and traditional knowledge towards global support for the conservation of biological diversity and sustainable use of its components.

REMARKS

The Nagoya Protocol constitutes the latest ambitious attempt to develop an international instrument complementing critical aspects of previous ABS instruments. With regard to access, some major achievements were reached: the binding nature, obligations to ensure legal certainty, the facilitation of non-commercial research and stronger involvement of local and indigenous communities. However, the Protocol also suffers from several shortcomings such as an overuse of qualifiers ('as appropriate', 'where applicable', 'as far as possible' and 'if available') and weak language ('endeavour', 'encourage', 'consider' and 'promote') in central provisions.

Internal waters and territorial sea

This section introduces the UNCLOS rights and obligations relevant for access to marine genetic resources located within the internal waters and territorial sea of a coastal state and compares these with ABS provisions.

UNCLOS

The territorial sovereignty of a coastal state extends beyond its land territory and internal waters to the territorial sea (2.1). The coastal state, therefore, has the authority to regulate all exploration and exploitation activities using marine genetic resources (Zewers, 2008, p.168).

In contrast to the internal waters, sovereignty within the territorial sea is qualified by the obligation to grant innocent passage to foreign vessels (17). Passage means continuous and expeditious navigation through the territorial sea with the possibility of incidental stopping and anchoring (18). Innocent passage means passage that is not prejudicial to the peace, good order or security of a coastal state (19.1). Passage is prejudicial to the peace, good order or security of a coastal state as soon a foreign vessel engages in any fishing, research and survey or other activities that do not have a direct bearing on passage (19.2(i)–(j) and (l)). This list is not exhaustive, and the coastal state has discretion in determining further activities that could render passage of foreign vessels non-innocent, such as exploitation activities of mineral resources (Graf Vitzthum, 2006, para. 123). With regard to innocent passage, the coastal state may adopt laws and regulations, especially concerning the prevention of infringement of fisheries laws as well as marine scientific and hydrographic surveys (21.1(e) and (g)).

In line with the above provisions, the coastal state has, in the exercise of its sovereignty, the exclusive right to regulate, authorize and conduct marine scientific research in its territorial sea. In addition, the conduct of marine scientific research by foreign vessels always requires the express consent of the coastal state (245).

These provisions suggest that the coastal state has the exclusive right to regulate any exploratory and scientific activities on marine genetic resources in its territorial sea. The requirement of express consent also indicates that any information on planned research activities in the territorial sea must be submitted to the coastal state in advance.

Arguably, a certain derogation of exclusive coastal state rights can be inferred from the obligation to promote international cooperation and to create favourable conditions for the conduct of marine scientific research (239, 242, 243). However, these obligations are very broad and they neither confer any concrete commitments to the conduct of marine scientific research nor impose restrictions on the coastal state in its exercise of sovereignty (Wegelein, 2005, p.145).

COMPARISON OF UNCLOS AND THE ABS REGIME

Comparing the rights and obligations of both conventions concerning commercial activities using marine genetic resources in the internal waters and territorial sea, states have unequal discretionary powers. While the coastal state is under no obligation to facilitate the exploration and exploitation of its natural resources under UNLCOS, a different picture emerges under the CBD. Here, a provider state 'shall endeavor to create conditions to facilitate access to genetic resources', which includes non-commercial and commercial uses (15.2 and 15.7). In addition, the Nagoya Protocol prescribes legislative, administrative and policy measures that support facilitated access. However, the wording is vague and does not create a specific obligation for provider states to grant access to genetic resources (Winter, 2009, p.20). This is underscored by Article 15.4 CBD prescribing that 'access, *where granted*, shall be on mutually agreed terms' (emphasis added), which underscores provider state powers to decide on access unilaterally. Although this interpretation indicates some consistency between UNCLOS and the CBD, the CBD's facilitation provision goes further than UNCLOS. The obligation 'not to impose restrictions that run counter to the objectives of [the CBD]' (15.2) suggests that access regulations must not impair the functioning of ABS, this being the third objective of the CBD (Wolfrum and Matz, 2000, p.469).

Although the CBD is more restrictive than UNCLOS, this does not imply that there is an unresolvable conflict between the conventions. A state being a party to UNCLOS enjoys full sovereignty. Sovereignty denotes the absolute power of self-determination of a state. It does not preclude self-restriction (Schermers, 2002, pp.186–187; Stemplowski, 2006, pp.239–240). Thus, a non-party to the CBD may voluntarily decide to restrict itself and facilitate access, while a party to the CBD must adhere to self-restriction and facilitate access. Neither case violates the concept of sovereignty under UNCLOS.

If an activity is declared as marine scientific research on marine genetic resources, both conventions provide very similar rights and obligations (Verhoosel, 1998, p.99; Proelß, 2009, p.60). Both promote non-commercial research and demand prior authorization before access to genetic resources or a marine scientific research project takes place. In addition, UNCLOS provisions on promoting marine scientific research and the CBD provision on facilitating access to genetic resources are equally weak, because neither requires specific obligations to be met.

Exclusive economic zone and the continental shelves

This part analyses UNCLOS provisions relevant to the access of marine genetic resources in the EEZ and continental shelf and compares these to the ABS system.

UNCLOS

In the EEZ, the coastal state has sovereign rights for the purpose of exploring, exploiting, conserving and managing the natural resources (including marine genetic resources), whether living or non-living, of the waters superadjacent to the seabed and its subsoil (56.1(a)) (Zewers, 2008, p.168). Sovereign rights over living resources within the EEZ must be read in conjunction with those provisions dealing with the conservation and optimum utilization of living resources, mainly Articles 61 and 62 but also 63–67 (Lagoni and Proelß, 2006, p.225). These provisions limit sovereign rights by obliging those coastal states that do not have the capacity to harvest their entire allowable catch of living resources in the EEZ 'to give other States access to the surplus of the allowable catch ...' (61.1 and 62.2). Because marine genetic resources belong to living resources under UNCLOS, this raises the question of whether this obligation constitutes a derogation of a state's sovereign right to determine access to its marine genetic resources.

An analysis of the language used for UNCLOS provisions on living resources in the EEZ shows that they do not constitute an obligation to provide access to marine genetic resources (Farrier and Tucker, 2001, pp.223–224). The use of 'harvest' and 'allowable catch' in correlation with 'conservation' and 'optimum utilization' immediately indicates the exploitation of a resource required in bulk amounts, whereas marine genetic resources are initially sampled in small amounts (maximum 1kg). Additional terms used in relation to the surplus of allowable catch, such as 'fishing', 'stocks', 'fishermen', 'fishing vessels', 'quotas of catch', 'landing of ... catch' (62.4), sufficiently indicate that access to the surplus was developed with commercial fishing for consumption in mind (Buck, 2007, p.214). Thus, the duty to provide access to the surplus of living resources does not apply to marine genetic resources and, as such, does not derogate state sovereign rights to control the exploration and exploitation of marine genetic resources. This also implies that the power of coastal states to control the commercial exploration and exploitation of marine genetic resources remains unqualified in the EEZ.

For the continental shelf, a coastal state exercises sovereign rights for the purpose of exploring and exploiting its natural resources (77.1). The natural resources of the continental shelf comprise mineral and non-living resources, as well as living organisms belonging to sedentary species, i.e., organisms which are at the harvestable stage, either immobile on or under the seabed or are unable to move apart from constant physical contact with the seabed (77.4). This includes many species that are mainly targeted as marine genetic resources, such as sponges, cold water corals, ascidians and snails. Sovereign rights over sedentary species on the continental shelf are also unqualified, because no other state may explore or exploit the natural resources without the express consent of the coastal state (77.2). Therefore, the control of the coastal state concerning the exploration and exploitation of the marine genetic resources of the continental shelf is also without restriction.

The legal situation is very different compared with marine scientific research in the EEZ and on the continental shelf. UNCLOS confers jurisdiction upon a coastal state to regulate, authorize and conduct marine scientific research in the EEZ and the continental shelf (246.1 and 56.1(b)(ii)). All marine scientific research activities in these areas require the consent of the coastal state (246.2), which is partly obligatory and partly subject to the discretionary powers of the coastal state.

A coastal state is obliged to grant its consent for marine scientific research projects in normal circumstances (obligatory consent) (246.3) (Hafner, 2006, para. 259). The meaning of 'normal circumstances' can be deduced from the same provision (Gorina-Ysern, 2003, p.315): normal circumstances apply when research activity is a) in accordance with UNCLOS (does not interfere with other legitimate uses of the sea or is harmful to the marine environment), b) exclusively for peaceful purposes and c) increasing scientific knowledge of the marine environment for the benefit of all mankind. The last point also suggests that marine scientific research, in normal circumstances, is not geared towards profit-generation (CBD SBSTTA, 2003a, para. 47). In these cases, consent may not be delayed or denied unreasonably.

However, a coastal state has the discretion to withhold its consent for a marine scientific research project (facultative consent) (Hafner, 2006, para. 259), if information on the research project is inaccurate or if the project is of direct significance for the exploration and exploitation of non-living or living natural resources (246.5(a)). To clarify this provision, states should seek to promote through international organizations, mainly the Intergovernmental Oceanographic Commission, the establishment of general criteria and guidelines to ascertain the nature of any marine scientific research (251). To date, no such criteria and guidelines have been unequivocally agreed upon, and it is mainly at the discretion of the coastal state to determine the nature of marine scientific research projects.

For a coastal state to determine the nature of a research project and thus whether or not to grant consent for it depends on a request for consent by the researching state or international organization. The request for consent must be submitted at least six months before the research activity will take place and must provide information on various aspects of the research project (248) (Soons, 1982, p.170). This includes (248(a)–(f)) (UNDOALOS, 2010):

- a full description of the nature and objectives of the project;
- the methods and means, including vessel details and description of scientific equipment;
- the precise geographical area in which the project will be conducted;
- the research dates from appearance to departure of the research vessel;
- the name of the sponsoring institution, its director and the person in charge of the project; and
- the extent to which the coastal state will participate or will be represented in the project.

The coastal state may request additional information (UNDOALOS, 2010), but this is balanced by the obligation to adopt reasonable rules, regulations and procedures to promote and facilitate marine scientific research beyond their territorial sea (255). Moreover, a coastal state provides implied consent if, within four months of receiving the request for consent, it does not inform the researching entity that it is withholding its consent, it requires more information or it is calling for outstanding obligations. In such cases, the research activity may begin six months after the initial request was made (252).

Finally, UNCLOS gives a coastal state the right to suspend or cease marine scientific research activities under certain circumstances (253). For example, a coastal state may suspend such activities when they do not conform to the information provided in the request for consent or do not comply with conditions set by the coastal state (253.1).

COMPARISON OF UNCLOS AND THE ABS REGIME

Before the rights and regulations of both conventions concerning activities on marine genetic resources within the EEZ and on the continental shelf can be compared, it must be noted that the existence of the EEZ, and hence the applicability of the ABS regime to this zone, depends on the express proclamation or declaration of the coastal state (Attard, 1987, pp.54–59 and 141–142). The practice of a majority of coastal states underscores the need to expressly proclaim an EEZ. Unless a coastal state has declared an EEZ, the waters superadjacent to the continental shelf belong to the high seas and are outside the ambit of the ABS system. It is emphasized here that the ABS provisions for the EEZ apply to a maximum distance of 200 nautical miles from the baseline, where this distance represents the minimum breadth of the continental shelf.

Under UNCLOS, the coastal state has all rights to control the commercial exploration and exploitation of marine genetic resources within its EEZ and continental shelf. The coastal state is under no obligation to provide or facilitate access to marine genetic resources to foreign nationals wishing to exploit such resources. Compared with the ABS regime of the CBD, a similar picture emerges to that which we saw with the territorial sea. Again, a state's right to determine access is unqualified under UNCLOS, whereas the CBD calls for at least facilitated access. The provisions of the CBD only affect these rights if the exercise of those rights would cause serious damage or pose a threat to biological diversity (22.1) – a situation that may occur when collecting marine genetic resources (Hunt and Vincent, 2006, pp.58 and 60). In such cases, CBD provisions, especially those on conservation and sustainable use, precede UNCLOS provisions, and users collecting marine genetic resources must abide to these as well.

With regard to marine scientific research, UNCLOS prescribes that the activity can only begin with the coastal state's consent. This accords with the requirement for prior informed consent under the CBD.

However, UNCLOS goes further than the CBD by also laying down specific information to be contained within the application for consent. As the CBD itself does not provide additional guidance in this regard, the Bonn Guidelines must be consulted. Like UNCLOS, the Bonn Guidelines prescribe that the application for prior informed consent includes information about the intent, methods, geography and dates of access as well as information about the applicant and possible participation. The Guidelines go even further than UNCLOS by calling for detailed benefit-sharing schemes in the application.

Although both conventions call for prior consent, the marine scientific research regime completely lacks the obligation for access to be on mutually agreed terms. As a result, it could happen that, during access negotiations, a coastal state abuses its discretion to withhold consent in order to push for its own priorities. Thus, the researching entity may be in the unfavourable position of having to comply with overly demanding conditions when negotiating access.

Concerning the facilitation of non-commercial research activities, both conventions contain provisions with similar effects. Under UNLCOS, coastal states must grant their consent in normal circumstances, i.e., for non-commercial purposes (246.3). Under the Nagoya Protocol, member states must adopt simplified measures on access for non-commercial research purposes (8(a)). However, as soon as a research project pursues economic aims, the discretion attributed to coastal states and provider states differ.

The central question is whether a coastal state's right to withhold consent for research of direct significance for (commercial) exploration and exploitation under UNCLOS provides more discretion to withhold consent than the CBD. Under the CBD, states must at least facilitate access, which has prompted legal authors to conclude that UNCLOS conveys stronger rights to decline access than the CBD (Henne, 1998, p.328; Wolfrum and Matz, 2000, pp.470 and 479).

However, the view that UNCLOS provides stronger rights to decline research projects than the CBD can be contested. First, under UNCLOS, a coastal state ceases to have sovereign rights and has only jurisdiction if an activity using marine genetic resources qualifies as marine scientific research (56.1(b)(ii)). Equipping a coastal state with only jurisdiction, that is the capacity 'to exercise, in conformity with international law, of legislative, executive and judicial functions over the sea and over persons and things on or under the sea' (Marston, 1989, p.316), suggests a regime weaker than that provided by the sovereignty or sovereign rights of coastal states (Brown, 1994, p.221; Gavouneli, 2007, p.64). While sovereignty and sovereign rights comprise almost full and exclusive jurisdiction, jurisdiction only entails the exercise of certain aspects of sovereignty (Wegelein, 2005, p.106). In other words, jurisdiction over marine scientific research indicates that the rights of other states are so numerous and complex that coastal state rights cannot be summarized under an exclusive concept such as sovereignty or sovereign rights (UNDOALOS, 2010).

Second, a coastal state has to investigate the nature of the research project in order to support its discretion to withhold consent. Therefore, the coastal state relies only on the information provided in advance by the researching entity. If the coastal state receives all relevant information, it must prove that the research project has *direct* significance for the exploration and exploitation of natural resources. This is not straightforward. The term 'direct' adds an important qualifying element, which does not allow the coastal state to regard all research as significant for the exploration and exploitation of natural resources. 'Direct' has been interpreted to mean that the data from marine scientific research projects must allow for exploration and exploitation in the near future with the technology currently available (Wegelein, 2005, p.87). 'Direct' does not include research that can become useful in the future when combined with other data yet to be collected (Soons, 1982, p.171). In contrast, the sovereign rights conferred by the CBD apply to any activity, regardless of its intent. The provider state is hardly obliged to prove whether or not an activity is of commercial significance before applying its discretionary powers to grant consent. A minor derogation can be found in Article 8 Nagoya Protocol, calling for simplified measures for non-commercial research. However, this derogation is qualified in the same article by the addition that states need to take 'into account the need to address a change of intent for such research'.

Third, UNCLOS obliges coastal states to

> endeavour to adopt reasonable rules, regulations and procedures to promote and facilitate marine scientific research conducted … beyond their territorial sea … and promote assistance to marine scientific research vessels which comply with the relevant provisions of [the marine scientific research] Part'.
>
> (255)

The formulation is broad and implies an obligation for all coastal states specifically regulating marine scientific research in their EEZ and continental shelf (Nordquist, 2002, p.602). In addition, this obligation applies regardless of the intent of marine scientific research, i.e., it also applies to research with a direct significance to exploration and exploitation (Soons, 1982, p.212).

Finally, despite the right of coastal states to regulate research and make research by foreign nationals dependent on their consent, the obligation to provide undelayed consent in 'normal circumstances' also degrades coastal state powers.

In conclusion, the CBD provides stronger rights to the provider state to determine access, than are provided by UNCLOS within its consent regime for marine scientific research in the EEZ and on the continental shelf.

That the CBD provides more discretion to determine access by research projects than UNCLOS could give rise to a potential conflict. If a coastal state, party to both UNCLOS and the CBD, exercises its sovereign rights under the CBD to prohibit marine scientific research on genetic resources, it is in breach of international law, because, under UNCLOS, it is under a strong obligation to grant consent to marine scientific research.

An analysis of the precedence of the conventions is helpful to ensure correct implementation and avoid this conflict. First, the CBD states that its provisions 'shall not affect the rights and obligations of any Contracting Party deriving from any existing international agreement, except where the exercise of those rights and obligations would cause a serious damage or threat to biological diversity' (22.1). From this, it can be concluded from that the obligation under UNCLOS to grant consent to a marine scientific research project remains unaffected by the CBD, unless that project would cause serious damage or a threat to biological diversity. In such cases, the provider state may make its consent dependent on compliance with provisions on conservation, sustainable use and ABS.

Second, and more specifically, the CBD obliges member states to 'implement this Convention with respect to the marine environment consistently with the rights and obligations of States under the law of the sea' (22.2). Moreover, UNCLOS prescribes that it does 'not alter the rights and obligations of States Parties which arise from other agreements compatible with [UNCLOS] and which do not affect the enjoyment by other States Parties of their rights or the performance of their obligations under [UNCLOS]' (311.2). These two provisions make it clear that the CBD and UNCLOS do not affect or alter the rights and obligations of other agreements except under certain circumstances, and that both conventions exist in parallel and complement each other (Wolfrum and Matz, 2000, p.476). However, if the rights and obligations of the CBD and other agreements are incompatible with UNCLOS, then these must be interpreted to comply with UNCLOS or not be applied at all (Matz, 2002, p.218) – in every case, UNCLOS prevails (Proelß, 2009, p.59). Thus, if an activity on marine genetic resources is declared as marine scientific research, the access provisions of the CBD favouring the provider state are overridden by the relevant UNCLOS provisions which strengthen the position of the researching state. This pertains to: a) the strong obligation for coastal states to grant consent for marine scientific research in normal circumstances and b) the weak discretionary powers to withhold consent when the research is of direct significance for exploration and exploitation. Consequently, states must include measures within their national ABS legislation which favour activities on marine genetic resources that qualify as marine scientific research with no direct significance for exploration and exploitation.

High seas and the Area

This section describes the high seas and the Area regimes within UNCLOS and analyses their applicability to activities concerning marine genetic resources.

THE HIGH SEAS

On the high seas, a coastal state ceases to have any exclusive geographical jurisdiction over the region or the resources found therein (89), but it retains exclusive jurisdiction over ships flying its flag (92). All states enjoy certain regulated freedoms on the high seas, which render the high seas open to all states and subject to the jurisdiction of none (87) (CBD SBSTTA, 2003b, para. 5). The freedoms of the high seas under UNCLOS 'comprise, *inter alia*, both for coastal and land-locked States: (a) freedom of navigation; (b) freedom of overflight; (c) freedom to lay submarine cables and pipelines, subject to Part VI; (d) freedom to construct artificial islands and other installations permitted under international law, subject to Part VI; (e) freedom of fishing, subject to the conditions laid down in section 2; (f) freedom of scientific research, subject to Parts VI and XIII' (87.1).

It is not initially clear if the exploration and exploitation of marine genetic resources also fall under the freedom of the high seas. The reference to freedom of fishing suggests only commercial fishing for consumption. Further, the language of the specific regulations on conservation and management of living resources on the high seas (116 and 118–119) is as restrictive to fisheries as are the provisions on living resources within the EEZ (61–62). Nevertheless, there is no provision potentially excluding marine genetic resources from the high seas regime. In addition, the wording of UNCLOS on activities that fall under the freedom of the high seas is not exhaustive, since Article 87.1 refers to the freedoms of the high seas to comprise, *inter alia*, certain activities. For example, another activity, which is not mentioned but could be considered free for all states on the high seas, is the dumping of waste (Wolfrum, 2006, para. 11). It is therefore concluded that the exploration and exploitation of marine genetic resources is a freedom of the high seas.

Marine scientific research is an explicit freedom of the high seas. All states have the right to conduct marine scientific research in the water column beyond the limits of the EEZ (257). However, such research is also constrained by the exercise of the freedom of other states (87.2). Furthermore, marine scientific research in the high seas is subject to the continental shelf and the marine scientific research regime. These limitations are particularly relevant where a coastal state has not claimed an EEZ, and the high seas thus overlie the continental shelf. In such cases, the freedom to exercise marine scientific research on the seabed and subsoil does not apply and falls under the consent regime laid down in UNCLOS (246). Where a coastal state has successfully claimed a continental shelf exceeding 200 nautical miles (76.4–76.10), it may only exercise its discretion to withhold consent within specific areas designated for exploration and exploitation (246.6) (Wegelein, 2005, p.207). Further limitations on marine scientific research in the high seas can be derived from the general provisions within UNCLOS (238–241).

Consequently, access to marine genetic resources within the high seas is open to all states, whether the activity qualifies as exploration and exploitation or marine scientific research. In addition, no state may hinder access to marine genetic resources by acting in any manner that affects the freedom of other states.

THE AREA

The Area and its resources are the common heritage of mankind (136). The resources of the Area are all 'solid, liquid or gaseous mineral resources *in situ* in the Area at or beneath the seabed, including polymetallic nodules' (133(a)). This formulation covers only mineral resources and excludes all living resources. Consequently, marine genetic resources found on the seabed, ocean floor or in the subsoil do not fall under the common heritage of mankind principle since they belong to natural living resources.

The question about the legal nature of genetic resources within the Area remains. Because the high seas comprise *all* parts of the sea that are not included in the territorial sea or EEZ, it can be concluded that such resources fall under the freedom of the high seas and are res communis (UN, 2006, para. 30; Proelß, 2009, p.63). However, this view has not yet been unanimously accepted internationally (UN, 2011, para. 332).

It is not clear though, if other articles of the Area regime apply to activities on marine genetic resources. UNCLOS defines the term 'activities in the Area' as those which are organized, conducted and controlled by the International Seabed Authority (ISA) (153.1 and 157.1). Activities in the Area cover only those 'activities of exploration for, and exploitation of, the resources of the Area' (1.1(3) and 134.2). Consequently, no article within the Area regime which deals exclusively with 'resources', 'activities within the Area', 'minerals' or any special type of metal, such as nickel, can be applied to activities on marine genetic resources. This includes almost all articles within the Area regime relevant to resource exploitation (136–142 and 144–155) and, therefore, exploration and exploitation activities on marine genetic resources fall completely outside the regulatory ambit of the Area regime (CBD SBSTTA, 2003b, para. 6). The only provision which could potentially be applied to activities on marine genetic resources is the regulation of marine scientific research within the Area.

Marine scientific research is distinct from exploration and exploitation in general and therefore does not fall under 'activity in the Area' (Brown, 1994, p.430; Gorina-Ysern, 2003, p.323). As marine scientific research is not defined for the Area regime, it can be assumed that the relevant provisions apply to scientific research on marine genetic resources as well. Under UNCLOS, all states and competent international organizations have the right to carry out marine scientific research, in conformity with the Area regime (256). According to the Area regime, marine scientific research is to be conducted for exclusively peaceful purposes and for the benefit of mankind as a whole, in conformity with the marine scientific research regime (143.1). This cross-reference points to the general provisions and international cooperation for marine scientific research in Articles 238–244. Because marine scientific research is not an 'activity in the Area', it also falls outside the regulatory power of the ISA (Wegelein, 2005, p.213). Nevertheless, the ISA itself may carry out marine scientific research (243.2).

COMPARISON OF UNCLOS AND THE ABS REGIME

The CBD does not empower a contracting party to decide unilaterally on access to marine genetic resources by other states in areas beyond the limits of national jurisdiction. This can be explained by the lack of territorial jurisdiction in such areas, but it is also the case that the jurisdictional scope of the CBD changes for areas beyond national jurisdiction. In these areas, the provisions of the CBD cease to apply to components of biological diversity. This implies that the ABS regime does not apply either, because its object of regulation, i.e., genetic resources, is a component of biological diversity. Beyond the limits of national jurisdiction, provisions of the CBD apply only for 'processes and activities ... carried out under [a Contracting Party's] jurisdiction or control ...' (4b). CBD provisions that deal with processes and activities are concerned with the identification and monitoring of processes and activities with adverse impacts on biological diversity (7(c)), their regulation and management (8(l)) and the impact assessment and minimization of such processes and activities (14(c) and (e)). These provisions confer only very broad and weak obligations on processes and activities that adversely affect components of biological diversity (CBD SBSTTA, 2003a, para. 70). They do not constitute any abridgement of a party's discretion to decide to access marine genetic resources in areas beyond national jurisdiction. As a result, access to marine genetic resources in areas beyond the limits of national jurisdiction is unregulated under the CBD; the ABS regime does not apply (CBD WG-ABS, 2005c, para. 29). Each contracting party is free to determine access for entities under its jurisdiction and control.

A similar picture emerges for the regulation of access to marine genetic resources under UNCLOS. In the high seas, access for the purposes of exploration and exploitation or marine scientific research falls under the freedom of the high seas and is therefore open to all states. Under the Area regime, access for the exploration and exploitation of marine genetic resources is completely unregulated, as in the CBD. Access for the purposes of marine scientific research is a right for all states.

THE WAY FORWARD

Although activities on marine genetic resources in areas beyond national jurisdiction are not completely unregulated, such as benefit-sharing provisions concerning marine scientific research in the Area (see below), existing provisions are fragmented and far from having established a functioning ABS system. The international community has acknowledged that such a virtual 'free-for-all' is unsatisfying from both a conservation and ethical perspective (Greiber, 2011, pp.7–8). From the conservation side, sampling events at fragile deep-sea ecosystems could become so numerous or destructive that biological diversity is threatened; in addition, the benefits potentially generated from utilization would be lost, instead of flowing to projects aimed at the conservation and sustainable use, such as marine protected areas in the high seas. Regulating the flow of benefits for such purposes could be part of a larger package dealing with conservation and sustainable use of marine biodiversity in general (IUCN and BfN, 2011, p.2). From an ethical or philosophical perspective, marine genetic resources could have a legal status similar to the common heritage of mankind: it would be the responsibility of entities with the capacity to access and use marine genetic resources to share the benefits with those that do not have that capacity. Also, access regulation in coastal states may impel access seekers to obtain their samples in the unregulated high seas or the Area.

Leaving aside the contentious discussion on the legal status of marine genetic resources in areas beyond national jurisdiction (Leary, 2012, p.437), various solutions are discussed in international forums. The main solutions for capturing activities on marine genetic resources in areas beyond national jurisdiction presented so far have involved either expanding the scope of existing mechanisms, such as the Area regime (Arico and Salpin, 2005, p.181; Pisupati *et al.*, 2008, p.60; Arnaud-Haond *et al.*, 2011, p.1522) or regional fisheries organizations, or developing new mechanisms, such as the global multilateral benefit-sharing mechanism as envisaged by the Nagoya Protocol (10) or an implementation agreement under UNCLOS (CBD SBSTTA, 2003a, paras 113–129; Proelß, 2009, pp.64–70; Greiber, 2011, pp.29–34 and 44–49). Each approach has its advantages and disadvantages.

Expanding the scope of the Area regime and the mandate of the ISA implies that the Authority would somehow determine and negotiate access to genetic resources, monitor utilization, deal with non-compliance and regulate (global) benefit sharing. This is advantageous in that it would avoid multiple agreements and organizations regulating activities in the same area. In addition, the Area regime envisages (global) benefit sharing, which is mirrored by the third objective of the CBD and the Nagoya Protocol. However, it is questionable whether the Area regime and the Authority are adequate for the regulation of all of these activities. First, the Authority has not yet passed any serious operational test (Proelß, 2009, p.66). Second, regulating all these activities might quickly outrun the Authority's capacities. To date, eight entities have contracts with the Authority to conduct activities in the Area. Activities concerning exploration and exploitation of marine genetic resources are much more numerous (CBD ABS GTLE, 2008b, p.9), and the Authority cannot organize and control all of these. Third, extending the Area regime would involve lengthy discussions with member states on how to integrate marine genetic resources into the existing regime. By way of comparison, negotiating the Area regime has lasted for the past 40 years and is still not fully concluded. To negotiate a similar regime for marine genetic resources would arguably be quicker, but would still take a lot of time.

As with the ISA approach above, regional fisheries organizations have the advantage that they are already existing instruments for regulating access to living resources in areas beyond national jurisdiction. Nevertheless, they also share many of the disadvantages of the ISA approach. First, they aim at the optimum generation of profits instead of conservation. Second, they regulate a completely different set of extractive methods and activities. Third, they use mainly quotas to regulate state behaviour, which is an inappropriate instrument to regulate access to genetic resources. Fourth, they do not handle intellectual property rights. Fifth, they are inexperienced in benefit sharing and the negotiation of ABS agreements.

Although the Nagoya Protocol does not mention areas beyond national jurisdiction, it does provide several opportunities for the international community to adopt a specialized ABS-agreement covering these areas (4.2 and 10). The main instrument is the 'global multilateral benefit-sharing mechanism' that addresses benefit sharing for those genetic resources for which it is not possible to grant or obtain prior informed consent (10). Genetic resources occurring in the high seas and the Area would clearly qualify as such resources. The benefits derived from the utilization of such resources should be used for the conservation of biological diversity and sustainable use globally. Although drafting an agreement on such a mechanism would be difficult in terms of avoiding conflicts with both the CBD and UNCLOS, it would provide several advantages. It would be a sound alternative to the shortcomings of already existing mechanisms, CBD forums are experienced in managing genetic resources, and such a mechanism could also address benefit sharing for transboundary genetic resources – an issue that has become the focus of international attention recently.

The drafting of an implementation agreement receives most attention by the UN Informal Working Group to Study Issues Relating to the Conservation and Sustainable Use of Marine Biological Diversity Beyond Areas of National Jurisdiction (UN, 2012). Such an agreement would complement existing mechanisms already managing the Area and high seas and would strive for conservation and sustainable use by establishing a regime on area-based management tools, environmental impact assessment, capacity building, transfer of technology and ABS. Current negotiations are still in their infancy, and future ones might use existing mechanisms and agreements, such as those mentioned above, as templates, while minding their advantages and limitations. However, an implementation agreement could also run into troubles. An implementation agreement could create a completely decentralized approach by only obliging flag states to share benefits. As each member state can interpret provisions differently, such an approach might create a fragmented system that insufficiently meets benefit-sharing targets. Users might choose to fly flags of convenience from states with undemanding conditions, thereby avoiding benefit-sharing obligations. One option to avoid this issue would be to oblige port states to adopt benefit-sharing provisions for ships calling at their ports. As changing shipping routes to port states with less demanding benefit-sharing obligations might involve costs exceeding benefits that have to be shared, this could provide an alternative means of ensuring a functioning system.

This very brief overview allows for the conclusion that all options are at comparably early stages in their development. However, two approaches have received particular attention in international discussions: a specialized agreement as envisaged by the Nagoya Protocol or an implementation agreement under the aegis of UNCLOS. Both options are part of a larger regulatory package dealing with biodiversity in areas beyond national jurisdiction in general. While the rationale for such an instrument has been clearly stated, many detailed issues remain unresolved. These involve access rules which balance conservation and use interests, the types of benefits to be shared and the beneficiaries, the level of monitoring and control, the establishment of central institutions and geographical scope. Although such an analysis is imperative, it is beyond the scope of this work.

Regulation of access to derivatives

A contentious issue for negotiations concerning ABS is that of 'derivatives' that originate from genetic resources. The main issues are: a) whether the ABS system and UNCLOS regulate access to derivatives or not and b) in case of ABS, how 'much' benefit sharing providers may claim during access negotiations from the use of derivatives.

DERIVATIVES UNDER THE CBD

The Nagoya Protocol defines the term 'derivative' as a 'naturally occurring biochemical compound resulting from the genetic expression or metabolism of biological or genetic resources, even if it does not contain functional units of heredity' (2(e)). In other words, derivatives are any biological molecules that originate from biological processes within organisms and include nucleic acids, proteins and other biological molecules. This definition leaves out other options for the interpretation of 'derivatives', which were proposed in earlier discussions (CBD WG-ABS, 2008, para. 20).

Several problems in relation to access to biological material are associated with the concept of derivatives. First, it is not clear whether the collection of derivatives also falls under access to genetic resources. Strictly speaking, access to derivatives would not qualify as access to genetic resources, if the derivatives do not contain functional units of heredity. In such cases, a provider state cannot negotiate ABS agreements. However, such situations will rarely be the case. Access often involves the taking of whole organisms or parts thereof that contain functional units of heredity (CBD COP, 1995d, para. 51). Apart from that, access can be detached from its combination with collection and extended to also include the intended use. From this perspective, access to genetic resources would take place when the collector of an organism intends to engage in 'utilization of genetic resources'. As utilization also covers derivatives without functional units of heredity, access to such derivatives would trigger benefit sharing, and the provider could negotiate benefit-sharing agreements. Moreover, excluding derivatives without functional units of heredity would weaken the incentive to ensure conservation and sustainable use, which in turn undermines the objectives of the CBD (CBD COP, 1996a, para. 33).

Second, it is a difficult question as to whether and how the national access legislation of provider states covers the transfer of products from the initial user to third parties. Typically, the user transferring products to third parties is in another country, and it is not possible to extend access legislation to areas that lie beyond a provider state's national jurisdiction (although it is still free to negotiate transfer conditions to third parties under a material transfer agreement) (Lesser, 1998, p.48). Furthermore, it would not be desirable to extend access legislation in this way, because it would be almost impossible to anticipate all the different circumstances in which products are transferred, unless the prior informed consent of the provider state was required each time a product was passed on (Glowka, 1997b, p.37). Compliance with such prescriptions would be practically impossible to monitor and control by the provider state (Dross and Wolff, 2005, p.60). The question then arises as to whether provider state legislation is the right instrument to control transfer to third parties, or whether there are other instruments on the user side that are better suited to fulfil this task (Chapter 5).

Third, it is not clear if the provider can negotiate benefits for envisaged end products that are too 'dissimilar' to the resource originally accessed. There are no clear criteria or guidelines that provide a definite answer. Generally speaking, all uses should trigger benefit sharing, no matter how dissimilar the end product may be. The only variable is the degree of benefit sharing a provider can claim, which depends on the relationship between a final product and the original sample (ten Kate and Laird, 2000b, p.67). For example, if the user is not able to synthesize the end product but has to return to the provider state for supply, the share of benefits is higher. Another example is the similarity of the end product. The more dissimilar, modified an end product is, the smaller the share of benefits becomes. No generally accepted criteria and guidelines exist here either, and it is up to the parties of ABS agreements to decide upon the degree of benefit sharing, based on questions of supply and the similarity of the end product to the original resource.

DERIVATIVES UNDER UNCLOS

The scope of UNCLOS provisions on derivatives is not initially clear. UNCLOS only refers to natural resources and does not explicitly discriminate between marine genetic resources and their derivatives.

With regard to commercial activities using natural resources, the only indication can be inferred from a coastal state's sovereign rights over the exploration *and exploitation* of natural resources within its waters. The broad reference to exploitation, i.e., economic utilization, suggests that a coastal state may regulate access to any marine genetic resources and their natural derivatives, notwithstanding their actual commercial utilization and the type of product resulting from utilization. It is important to state that conditions imposed on utilization, such as restrictions on the forwarding of intermediary or final products to third parties, should be negotiated during access if they are to be made at all. Once the resource has left the country, it is beyond the jurisdictional scope of the coastal state.

For marine scientific research, the coastal state has the general right to regulate and authorize any research project on natural resources within its waters. Specifically, the consent regime recognizes a relationship between the research object and any derivatives arising thereof (Gorina-Ysern, 2003, pp.346 and 355). For this purpose, a coastal state may make its consent dependent on certain conditions (Wegelein, 2005, p.188). Conditions might include, for example, that the coastal state is provided with access to the final results, conclusions, data and samples arising from marine scientific research (249(1)(a)–(g)). While this list is exhaustive for marine scientific research in normal circumstances (Soons, 1982, p.188), a coastal state may impose any condition it deems fit for marine scientific research with direct significance towards exploration and exploitation of natural resources (249(2)).

Sharing of benefits from utilization of marine genetic resources

The utilization of marine genetic resources may yield products which deliver a broad spectrum of monetary and non-monetary benefits (Reid, 1994, pp.241–266). Users tend to take the position that their monopoly over retaining benefits from commercialization is justified in the light of their high investment in research and development. Providers contest this by arguing that they are the titleholders of genetic resources (Francioni, 2006, p.21). In international law, UNCLOS and the CBD provide regulations that elaborate at length how any benefits derived from utilization of marine genetic resources must be shared between the user and provider side.

General benefit-sharing provisions under UNCLOS

In UNCLOS, the regimes on marine scientific research, as well as development and transfer of technology, contain provisions relevant to the sharing of benefits derived from the utilization of marine genetic resources (CBD WG-ABS, 2009, p.16).

With regard to marine scientific research, states or international organizations have to publicize and disseminate any knowledge resulting from research through appropriate channels as long the disclosure is not contrary to essential security interests (244.1 and 302). In addition, states should share benefits through the flow of scientific data and information, the transfer of knowledge and the strengthening of research capabilities through the training of scientific personnel (244.2).

SHARING BENEFITS THROUGH THE TRANSFER OF TECHNOLOGY

The conduct of marine scientific research requires marine technology. However 'marine technology' is not defined within UNCLOS but in guidelines developed by the Intergovernmental Oceanographic Commission, being a competent international organization under UNCLOS (271). It comprises the instruments, equipment, vessels, processes and methodologies that are necessary to study and understand the nature and resources of the marine environment (IOC, 2005, p.7). This includes: information and data; manuals, guidelines, criteria and standards; sampling and methodology equipment for water, geological, biological or chemical samples; observation facilities for remote sensing, buoys and tide gauges; equipment for in situ and laboratory observation; computers and software and expertise, knowledge, skills, as well as scientific and legal understanding related to marine scientific research.

UNCLOS establishes various principles and provisions on the transfer of marine technology which are relevant for benefit sharing. States are required to cooperate to promote the development and transfer of marine science and technology on fair and reasonable terms (266.1). More specifically, in order to accelerate social and economic development, states are obliged to promote the improvement of marine scientific and technological capacity in developing states with regard to the exploration and exploitation of marine resources as well as marine scientific research (266.2). Under article 267 UNCLOS, states' interests are protected insofar that states shall have due regard for the 'rights and duties of holders, suppliers and recipients of marine technology.' Thus holders of intellectual property rights are not restrained in their protected rights (Hafner, 2006, pp.453–454).

In addition, UNCLOS mentions basic objectives that underlie the transfer of marine technology. These are the dissemination of and facilitated access to technological knowledge, the development of technological infrastructure to facilitate transfer, the development of human resources through training and education and international cooperation at all levels (268). The means to achieve these objectives are programmes for the effective transfer of marine technology to states that need technical assistance for developing their own technological capacity for marine science and exploitation of resources (269(a)). Other means are conferences, seminars and symposiums; the exchange of scientists and experts and joint ventures (269(c)–(e)).

General benefit-sharing provisions under the ABS regime

The CBD prescribes the particularities of benefit sharing within the Convention text, the Bonn Guidelines and the Nagoya Protocol.

CONVENTION TEXT

The sharing of benefits arising from the utilization of genetic resources is one of the main objectives of the CBD (1) (Suneetha and Pisupati, 2009). The exact types of benefits are negotiated on mutually agreed terms, which indicates that particularities of each ABS transaction are for parties to decide on thereby creating a higher degree of flexibility towards the implementation of benefit sharing (CBD COP, 1996b, p.7; Wolfrum, 1996, p.387). Benefit sharing is then triggered by prior informed consent and subsequent utilization (CBD COP, 1998c, para16).

The CBD text identifies various types of benefits (Stoll, 2004, p.79). First, the CBD recognizes that the utilization of indigenous and local knowledge concerning genetic resources yields benefits which must be shared with the holders of such knowledge (8(j)). Second, user states should endeavour to conduct research on genetic resources within the provider state and with the full participation of the provider state (15.6). More specifically, the CBD calls for the effective participation of provider states in biotechnological research activities on its genetic resources (19.1). Third, the Convention envisages the fair and equitable[11] sharing of results from the research and development of genetic resources as well as benefits from commercial and other utilization with the provider state (15.7). Similarly, states are obliged to promote and advance access by provider states, on a fair and equitable basis, to results and benefits arising from biotechnologies (19.2). Fourth, the CBD also regulates the transfer of technology (16). Mechanisms to transfer technology can be key-turning projects, foreign direct investment, joint ventures (equity capital brought in by those sharing an enterprise), licensing (right to use propriety technology in exchange for a fee), technical-service arrangements, joint R&D arrangements, training, information exchange, sales contracts and management contracts (CBD SBSTTA, 1996b, paras 39–43). The text specifically prescribes that provider states should be provided with access to and transfer of technology which makes use of genetic resources, which also includes technology protected by intellectual property rights (16.3 and 16.5). Finally, the CBD contains additional provisions that indirectly link utilization to benefit sharing, e.g., those provisions on the exchange of information, which prescribe a facilitated exchange of information in combination with technologies that make use of genetic resources (17.2, 17.1, 16.1), and those on technical and scientific cooperation, stipulating cooperation for the development and use of technologies (18.4 and 1) as well as joint research programmes (18.5 and 1) relevant for benefit sharing.

Beneficiaries are primarily the contracting parties providing genetic resources. Such parties are either countries of origin or those parties that acquired genetic resources in accordance with the CBD (15.3). These parties are free to determine further beneficiaries on a sub-national level, such as: governmental departments at national, regional or local levels; local or indigenous communities; owners, holders and administrators of land or sea areas and universities, research centres, non-governmental organizations, etc. (CBD COP, 1998c, para. 18).

As the benefit sharing provisions were often criticized for their vagueness and broadness, the Bonn Guidelines deliver more substance.

BONN GUIDELINES

With the aim of assisting the development of legislative measures to implement ABS, the Bonn Guidelines provide a non-exhaustive list of monetary and non-monetary benefits that can be shared (Appendix II) (CBD COP, 1996b, p.4).

Monetary benefits include:

- access fees to single samples;
- up-front payments, i.e., before access and utilization take place;
- milestone payments, which may vary according to the progress of development of commercial results or products (ten Kate and Laird, 2000b, p.66);
- royalty payments during the commercialization of final results or products;[12]
- licence fees;
- special fees;
- salaries;
- research funding;
- joint ventures; and
- joint ownership of intellectual property rights.

Non-monetary benefits include:

- sharing of research and development results;
- cooperation in (biotechnological) research activities as well as education and training;
- participation in product development;
- admission to databases and ex-situ facilities (CBD WG-ABS, 2008, p.15);
- transfer of knowledge and technology and strengthening capacities for technology transfer;
- institutional capacity building and resources to strengthen administrative and enforcement capacities related to access regulation;
- training related to genetic resources;access to scientific information relevant to biodiversity conservation;
- contributions to the local economy;
- research towards priority needs (health, food and security sectors);
- establishment of professional relationships and future collaborative activities;
- social recognition; and
- joint ownership of intellectual property rights.

The balance among benefits to be shared is determined on a case-by-case basis (para. 47). The benefits should be shared fairly and equitably between those parties which have contributed to management, scientific and commercial processes concerning the genetic resource (48).

Each party has to take legislative, administrative or policy measures to ensure that the benefits arising from the utilization of genetic resources, as well as any subsequent application and commercialization, are shared fairly and equitably with the provider state (Articles 5.1 and 5.5). The benefits as listed under the Protocol include monetary and non-monetary benefits and are almost verbatim the benefits listed in the Bonn Guidelines (5.4 and the Annex). In addition, the Protocol prescribes collaboration and cooperation in technical and scientific research and development (R&D) programmes, which preferably take place in, and with the participation of, provider states (23). In this regard, access to technology by, and transfer of technology to, developing country parties should be encouraged. The basic paradigm that conservation and sustainable use are both stimulated by maintaining potential of discovering valuable genetic resources is now explicitly complemented by the obligation to encourage the flow of benefits towards conservation and sustainable use (9).

Finally, the Protocol introduces extensive measures on improving capacities (22). Capacity is one of the core benefits under the Protocol (2(g)–(j)), and parties must cooperate in capacity-building, capacity development and the strengthening of human resources and institutional capacities (22.1). Therefore, developing country parties should conduct capacity self-assessments to identify their national needs and priorities (22.3). Key areas identified by the Protocol that require capacity-building include the understanding and implementation of the Protocol, negotiation of mutually agreed terms, development and enforcement of domestic legislation and endogenous research capabilities (22.4 and 22.5).

Benefit-sharing provisions in distinct maritime zones

This section introduces the rights and obligations provided by UNCLOS relevant to sharing benefits.

Generally, coastal states can, in the exercise of their sovereignty or sovereign rights on natural living resources, link the exploration, exploitation and marine scientific research relating to marine genetic resources to various conditions, such as the sharing of benefits. Specific provisions for distinct maritime zones with benefit-sharing effects can be derived from the marine scientific research regime within the territorial sea, the EEZ and continental shelf and the Area.

In the territorial sea, the coastal state may, in the exercise of its sovereignty, set conditions for undertaking marine scientific research which also cover benefit sharing (245). Although the nature of benefit sharing is not clear under UNCLOS, it can be assumed that it primarily deals with non-commercial benefits, owing to the non-commercial nature of marine scientific research.

In the EEZ and on the continental shelf, a researching state must comply with certain conditions. The researching state must ensure that the coastal state can participate, or is at least represented, in the research project and receives the final results and conclusions after completion of the research. Moreover, the researching state must: provide access to all data and samples; provide assessments of data, samples and results and ensure that research results are internationally available (249.1). These conditions only apply for marine scientific research in 'normal circumstances' and are without prejudice to conditions established by the coastal state, if research is of direct significance for the exploration and exploitation of natural resources (249.2). In such cases, the coastal state may impose any conditions it deems fit, for example, prior agreement before research results are made internationally available (Soons, 1982, p.192). Nevertheless, the coastal state is still bound to promote and facilitate marine scientific research beyond the territorial sea and cannot therefore demand conditions that effectively prohibit the undertaking of any marine scientific research (255).

In the Area, marine scientific research must be conducted for the benefit of mankind as a whole and in accordance with Part XIII UNCLOS on marine scientific research (143). One indication of what is meant by 'benefit of mankind' can be derived from article 140 UNCLOS. This article states that activities in the Area shall be carried out for the benefit of mankind, taking into particular consideration the 'interests and needs of developing states'. The reference to Part XIII indicates that the general provisions on marine scientific research, including those with benefit-sharing effects (244), also apply to scientific research within the Area. This also includes the dissemination of research results, the strengthening of research capabilities and the training of personnel through international research programmes (143.2–143.3). Although the Area regime provides for transfer of technology under article 144 UNCLOS as well, this article only refers to technology and scientific knowledge relating to 'activities in the area'. As 'activities in the area' concern activities on mineral resources only, this article cannot be used to derive any obligations towards transfer of technology on marine genetic resources.

Comparison of UNCLOS and the ABS regime

Both conventions prescribe that entities which collect and utilize marine genetic resources should share benefits with the state providing such resources (Pavoni, 2006, footnote 48). However, the conventions differ in the scope of benefits to be shared.

First, while the CBD expressly recognizes holders of indigenous and local knowledge to be eligible for benefit sharing, UNCLOS does not contain an analogous provision. Thus, if a state has ratified only UNCLOS, holders of knowledge on the utilization of marine genetic resources may not necessarily benefit if their knowledge is used. The potential number of holders affected by this lacuna is likely to be minuscule because: a) virtually all states have ratified the CBD, and b) indigenous knowledge concerning utilization of marine genetic resources is less relevant than it is in the terrestrial context and it is also likely to diminish the further offshore resource collection occurs (Greer and Harvey, 2004, pp.36, 38, 51; Owens and Chambers, 2004, p.33; Ridgeway, 2009, p.317).

Second, the CBD envisages that provider states receive a share of both monetary and non-monetary benefits for the collection and utilization of their genetic resources, whereas UNCLOS refers only to the sharing of non-monetary benefits in its regimes on marine scientific research and transfer of technology (CBD WG-ABS, 2009, p.16). Nevertheless, UNCLOS does not prohibit a coastal state from adopting additional conditions on monetary benefit sharing, as long as the undertaking of marine scientific research is promoted and facilitated.

The types of non-monetary benefits under both conventions, such as joint research, training and technology transfer, overlap substantially. This overlap underscores the importance of non-monetary benefits because they constitute a relatively fast way of enhancing a provider state's scientific and technological capabilities – a process that would otherwise consume large investments in time and money (CBD WG-ABS, 2001, p.109). Enhancing capabilities also enables a provider country to conduct its own value-adding research on genetic resources. Access to the results of value-adding research often yields higher revenues on the international market than does access to the raw material (UNCTAD, 2004, p.78).

Finally, benefit sharing under the CBD is primarily linked to the access to and utilization of a particular genetic resource (1). That means that states are only obliged to share benefits if they actually engage in collection and utilization. A different situation may emerge under the transfer of technology regime under UNCLOS. The obligation to transfer technology seems to be a general obligation on states, detached from specific activities of a researching state.

Monitoring, compliance and dispute settlement

Monitoring, ensuring compliance and a dispute settlement mechanism are essential tools for a functioning ABS system. Therefore, this section analyses how these are addressed under UNCLOS and the ABS regime.

General provisions under UNCLOS

UNCLOS does not introduce stand-alone monitoring and compliance regimes but contains relevant provisions in those parts which deal with distinct maritime zones. They are analysed below. However, the settlement of disputes is regulated generally by Part XV of UNCLOS.

UNCLOS prescribes a complex dispute settlement mechanism in three sections (Klein, 2005). The first section lays down general provisions. These oblige parties to settle disputes by peaceful means chosen individually by those involved in the dispute or under specific provisions of the UN Charter (279–280). The UN Charter generally obliges parties to seek resolution by negotiation, enquiry, mediation, conciliation, arbitration and judicial settlement (33.1), but it does not specify any procedures. Compulsory dispute settlement, as laid down in section 2 of UNCLOS, only applies if parties cannot settle their dispute by these means (286).

Compulsory dispute settlement is the central part of the whole UNCLOS dispute settlement mechanism. It offers four procedures that parties to a dispute may choose from (287.1): the International Tribunal for the Law of the Sea, the International Court of Justice, arbitration (Annex VII) or special arbitration (Annex VIII).

Compulsory dispute settlement is heavily qualified in section 3. As soon the dispute relates to a coastal state's discretionary powers to regulate exploitation and marine scientific research, it is subject to a special regime. The primary means to settle the dispute is still that of compulsory dispute settlement, the only exception being that the coastal state 'shall not be obliged to accept submission to such settlement' (297.2(a) and 297.3(a)). There exists only the obligation for compulsory dispute settlement when a coastal state 'has acted in contravention … to freedoms and rights of navigation, overflight or the laying of submarine cables and pipelines' or 'has acted in contravention … [to] the protection and preservation of the marine environment' (297.1) (Boyle, 1997, p.42; Treves, 1999, p.7).

With regard to exploitation in particular, UNLCOS only talks about 'fisheries' (297.3(a)). However, UNCLOS bases this exemption on the sovereign rights relating to living resources in general, which indicates that coastal states are also exempt from compulsory dispute settlement when the exploitation of marine genetic resources is involved.

In the case of marine scientific research, coastal states are freed from compulsory dispute settlement if the dispute relates to a coastal state's discretionary powers in the EEZ or its right to suspend and cease scientific research (246 and 253). In such events, the dispute should be submitted to a conciliation commission (297.2(b)). The commission is bound insofar as it may not question the power of a coastal state to withhold consent if a project is of direct significance to the exploration and exploitation of living resources.

These exemptions substantially weaken the powers of researching states to institute legal proceedings against a coastal state with regard to the exploitation and research of marine genetic resources.

General provisions under the ABS regime

The CBD specifies features of monitoring, compliance and dispute settlement within the Convention text, the Bonn Guidelines and the Nagoya Protocol.

CONVENTION TEXT

The provisions on monitoring in the CBD are mostly unrelated to the utilization of genetic resources. They concern only the monitoring of components of biological diversity relevant for in-situ and ex-situ conservation and sustainable use and the identification of activities with significant adverse impacts on conservation and sustainable use (7, 8–10). The utilization of traditional knowledge is an integral part of in-situ conservation, so a certain obligation to monitor the utilization of such knowledge can be concluded. This obligation is very weak, because the monitoring duty applies only to activities that are likely to have significant and adverse impacts on biological diversity. This would hardly be the case for activities on (marine) genetic resources, because most are conducted in laboratories (Hunt and Vincent, 2006, p.58).

To ensure compliance, the CBD only proposes incentive measures for the conservation and sustainable use of biological diversity (11). These include economically and socially sound measures. The consequences for incidents of non-compliance are not mentioned.

The CBD regulates disputes concerning the interpretation and application of the Convention only generally. Where a dispute between parties arises, they must seek resolution by negotiation first (27.1). If negotiations prove unsuccessful, they may request mediation by a third party (27.2). Only if these two mechanisms do not resolve the dispute can parties submit the dispute either to the International Court of Justice or for arbitration (27.3 and Annex II Part 1). Otherwise, the dispute must be submitted for conciliation (27.4 and Annex II Part 2).

BONN GUIDELINES

The Bonn Guidelines establish monitoring mechanisms for genetic resources but fail to elaborate them clearly and in detail. On the provider side, they entrust competent national authorities to monitor ABS agreements (14(c)). For this purpose, national authorities should maintain national registries that record all issued permits and licences (39). On the user side, the Guidelines oblige users to maintain all relevant data regarding genetic resources, including documents attesting prior informed consent, the origin, use and any benefits arising from the use (16(b)(vi)). Generally, the Guidelines propose that states should maintain national monitoring systems through the involvement of relevant stakeholders (55). Finally, it calls for states to adopt 'voluntary means of verification', without further explaining these (57–58).

With regard to compliance, the Guidelines put the main responsibility on the user. Users are responsible for ensuring compliance, and user states should support compliance with prior informed consent and mutually agreed terms (54 and 16(d)). In cases of violation, parties should decide on appropriate and proportionate remedies (61). These could include sanctions such as penalty fees (60).

The Guidelines almost completely neglect to mention dispute settlement. They only subordinate any settlement to the terms mutually agreed upon in initial contractual agreements (59).

NAGOYA PROTOCOL

The Protocol provides several novel approaches to monitoring compliance, with rules on the utilization of genetic resources. The most prominent approach is the use of 'checkpoints' designated by each party (17(a)). Although parties are free to designate checkpoints, they should be relevant to utilization or collection of information on utilization of genetic resources, e.g., to any stage of research, development, innovation, pre-commercialization or commercialization (17.1(a)(iv)). They could include research institutions, patent offices, regulatory agencies or even universities (Kamau *et al.*, 2010, p.256). Checkpoints require users to submit information on prior informed consent, mutually agreed terms, the source and the utilization of the genetic resource (17.1(a)(i)–(ii)). That information is then submitted to the relevant authorities, the provider party and the ABS Clearing-House Mechanism (CHM) (17.1(a)(iii)), which is responsible for sharing and updating of information on national ABS measures, focal points, relevant authorities, permits, best practice, etc. (14). Other monitoring mechanisms include the sharing of information through reporting requirements, cost-effective communication tools and systems and a mandatory, internationally-recognized certificate of compliance (17.1(b)–(c) and 17.2).

A certificate of compliance is basically a permit made available to the ABS CHM, containing non-confidential information about the issuing authority, date, provider, person to whom prior informed consent was granted, genetic resources, use, a unique identifier and a confirmation that prior informed consent was obtained and mutually agreed terms have been established (17.2–17.4). Finally, it is the obligation of each party to monitor the implementation of the Protocol and report regularly to the COP (29).

Despite the above measures to monitor the utilization of genetic resources, there are no comparable monitoring mechanisms mentioning explicitly the utilization of traditional knowledge under the Protocol. Only article 17.4(g)–(i) on prior informed consent and mutually agreed terms can be interpreted to not only include providers and users, but also communities holding genetic resources and traditional knowledge. Given the clear distinction the Protocol draws between the utilization of genetic resources and the utilization of traditional knowledge, this could constitute an omission with far-reaching consequences.

The Protocol contains provisions on compliance but leaves it primarily to the parties involved to decide on appropriate, effective and proportionate measures to ensure that genetic resources and traditional knowledge have been accessed in accordance with prior informed consent and that mutually agreed terms have been established (15.1 and 16.1). Similarly elusive are provisions addressing situations of non-compliance (15.2 and 16.2), and in cases of alleged violations, parties have a weak obligation to cooperate (15.3 and 16.3). As a consequence, subsequent experts' meetings will address these shortcomings and will consider and approve cooperative procedures and institutional mechanisms to promote compliance with the Protocol (30). Draft procedures and mechanisms have been developed by experts' groups of the Intergovernmental Committee for the Nagoya Protocol (ICNP), which currently negotiate the specific functions of a Compliance Committee, which promotes compliance and deals with cases of non-compliance (CBD ICNP, 2012b).

With regard to the resolution of future disputes, before access takes place, parties are encouraged to agree on the jurisdiction to which disputes would be submitted, applicable laws and options for alternative resolution, such as mediation or arbitration (18.1). In addition, parties should take effective measures on access to justice and mutual recognition and enforcement of foreign judgments and arbitral awards (18.3).

Provisions in distinct maritime zones

UNCLOS contains only provisions on monitoring and ensuring compliance within the EEZ and on the continental shelf.

A coastal state has a strong right to board and inspect any vessels in the exercise of its sovereign rights to explore, exploit and manage living resources (73.1). With regard to marine scientific research, the foreign research state is obliged to submit reports to the coastal state on the results of the research project (249.1(b)), allow it access to all data from the project (249.1(c)) and inform it about any major changes (249.1(f)).

In cases of non-compliance, the coastal state may arrest those conducting the research and seek judicial proceedings to ensure compliance regarding the exploration and exploitation of living resources (73.1). If research activity within the EEZ or on the continental shelf is not in line with the information given upfront or does not comply with conditions set by the coastal state, the coastal state has the right to suspend or even cease the activity, depending on the severity of the offence (253).

Comparison of UNCLOS and the ABS regime

UNCLOS and the CBD employ different approaches for effective monitoring, compliance and dispute settlement.

Under UNCLOS, monitoring is mainly the duty of the coastal state. The rights conferred by UNLCOS are very strong with regard to the exploitation of marine genetic resources but weaken as soon the activity qualifies as marine scientific research. The CBD pursues a more balanced approach by prescribing duties on both sides with a slight emphasis on the user side. In addition, the CBD establishes central information nodes.

A similar picture emerges with compliance. Under UNCLOS, the coastal state is in a strong position to enforce its rights, at least as long the activity occurs within its waters. The CBD provides little guidance and obliges parties on both sides to elaborate and adopt adequate measures – at least until an appropriate mechanism has been developed under the CBD.

With dispute settlement, the decisions taken by the coastal state can barely be even challenged under UNCLOS. The CBD adopts a different approach and leaves it primarily to the parties of the ABS agreement to decide on appropriate dispute settlement mechanisms.

Remarks

The above analyses have concentrated only on basic issues, and a detailed comparison of the ABS regime and UNCLOS can be found in Table 3.1.

In broad terms, the conventions have many similarities. First, both assure a state's regulatory power in the territorial sea, the EEZ and the continental shelf, which then provides the legal basis to regulate access to its marine genetic resources (Wolfrum and Matz, 2000, p.469). Second, with regard to access with commercial intentions, both conventions grant similarly strong rights to the provider state to regulate access in its waters. Third, regarding access for non-commercial purposes, both conventions call for less stringent access regulations. Fourth, both prescribe the sharing of similar non-commercial benefits. In these respects, the conventions support each other.

However, there are also a number of differences between the ABS regime and UNCLOS. First, UNCLOS does not consider the role of traditional knowledge. Second, UNCLOS and the CBD differ in the exact degree of discretionary powers for the regulation of access. Third, UNCLOS does not prescribe mutually agreed terms. Fourth, UNCLOS does not stipulate the sharing of commercial benefits. Finally, the conventions apply different monitoring, compliance and dispute settlement measures. These discrepancies do not necessarily lead to conflicts. On the contrary, each convention might fill a legal gap in the other and therefore they complement each other (Wolfrum and Matz, 2000, p.472; Gorina-Ysern and Jones, 2006, pp.224 and 280; CBD WG-ABS, 2009, p.16).

An analysis of national legislation will show how states have solved the dual applicability of UNCLOS and the CBD for activities on marine genetic resources.

National regimes on management of marine genetic resources

In order to illustrate the implementation of international provisions with regard to access to and the utilization of marine genetic resources, this section provides an analysis of relevant legislation within selected countries on ABS, exploration and exploitation of marine resources or marine scientific research. Legislation has been mainly retrieved from CBD (2012), UNDOALOS (2012) and IOC (2012). The emphasis of the analysis lies on the legislation of 'typical' user states (industrialized states), because they have a reputation for lagging behind in the adoption of ABS measures. The aims of this analysis are threefold. First, for a globally functioning ABS system it is important that user states adopt measures which oblige users under their jurisdiction to comply with foreign ABS provisions, such as prior informed consent and mutually agreed terms. It is therefore the aim to identify whether user states only regulate access to their genetic resources, or whether they also regulate their users when utilizing genetic resources obtained abroad. Second, since this work focuses on marine genetic resources in particular, any legislation on the exploration and exploitation of marine living resources and marine scientific research will be identified as well. Third, elements of national legislation that could be used as model legislation will also be examined.

Australian ABS legislation

Australia is a megadiverse country, rich in both terrestrial and marine genetic resources (Kriwoken, 1996, p.114; Department of the Environment and Heritage, 2005, p.1). The commercial potential of products from genetic resources is reflected in the National Strategy for the Conservation of Australia's Biological Diversity which ensures 'that the social and economic benefits of the use of genetic material and products derived from Australia's biological diversity accrue to

Table 3.1 ABS-relevant elements within international agreements (CBD, Bonn Guidelines, Nagoya Protocol and UNCLOS). Relevant articles or paragraphs (Bonn Guidelines) in parentheses. Acronyms: ABS: Access and benefit sharing; CBD: Convention on Biological Diversity; CHM: Clearing-house mechanism; CNA: Competent National Authority; CS: Continental shelf; EEZ: Exclusive economic zone; GR: Genetic resources; ICJ: International Court of Justice; IPR: Intellectual property rights; ITLOS: International Tribunal for the Law of the Sea; MAT: Mutually Agreed Terms; PIC: Prior informed consent; MTA: Material transfer agreement; TK: Traditional Knowledge and UNCLOS: United Nations Convention on the Law of the Sea

Elements	CBD	Bonn Guidelines	Nagoya Protocol	UNCLOS
Objectives	Conservation, sustainable use, fair and equitable sharing of benefits (Article 1)	Thirteen objectives (Paragraph 11)	Fair and equitable sharing of benefits contributing to conservation and sustainable use (Article 1)	Legal order for the seas, peaceful uses, equitable and efficient use of resources, study and preservation of marine environment (Fourth preamble)
Scope	Components of biological diversity, processes and activities under its jurisdiction (4)	Genetic resources; traditional knowledge, innovations, practices; benefits (9)	Genetic resources, associated traditional knowledge, benefits from their utilization (3)	State parties (Article 1.2(1)), self-governing associated states and territories, and international organizations (1.2(2), 305.1)
Legal status of genetic/natural resources	Sovereign rights (fourth preamble, 3, 15.1)	Sovereign rights (4, 24, 44(c))	Sovereign rights (third preamble, 6)	Territorial sea: sovereignty (2.1); EEZ, CS: sovereign rights (56.1, 77); High seas and the Area: res communis (87.1)
Facilitation of access	Facilitate access and not impose restrictions counter CBD objectives (15.2)	Facilitate access (24), Providers to avoid arbitrary restrictions on access (16(c)(ii))	Legal certainty, clarity, transparency; fair, non-arbitrary access rules; information on PIC applications; clear and transparent CNA decisions; clear rules and procedures on MATs (6.3(a)–(c) and (g))	Access to surplus of allowable catch (fisheries only; 62), promote and facilitate marine scientific research (239, 255), favourable conditions for research (243), consent in normal circumstances (246.3), implied consent (252)
Prior informed consent	Access subject to PIC (15.5)	Access subject to PIC (24), Basic principles and elements of a PIC system (26–27).	Access subject to PIC (6.1); Legislative, administrative, policy measures for PIC aiming at facilitated access (6.3), Criteria and/or processes for obtaining PIC from communities (6.3(f))	(Express) consent for marine scientific research required (245, 246.2); Duty to provide information (248); Prior agreement before disclosure of research results (249.2)

Elements	CBD	Bonn Guidelines	Nagoya Protocol	UNCLOS
Permit/certificate	–	Written form, e.g., a permit or licence (38–39)	Permit or equivalent verifying PIC and MATs (6.3(e)), international certificate of compliance, when permit submitted to ABS-CHM (17.2–3)	–
Mutually agreed terms	MAT on access (15.4), benefit-sharing (15.7), technology transfer (16.3), joint research programmes (18.5), access to results and benefits from biotechnologies (19.2)	MAT to ensure fair and equitable BS (41), Basic requirements (42), Guiding parameters (43), Indicative list of MAT (44)	MAT include dispute settlement, benefit-sharing, third party use, changes in intent, information on implementing MAT (5.1–2), 6.3(g)(i)–(iv), 17.1(b)); Clear rules and procedures for MAT and MAT in written form (6.3(g)), MAT with communities (5.5 and 7)	–
MAT parties	Implicitly: users and provider parties (15.7), parties to the CBD (18.5, 19.2)	Implicitly based on para. 35: competent national authorities (28), stakeholders (30), communities (31), and parties and stakeholders in benefit-sharing (41)	Providers and users of GR and TK (Preamble 10, 17.1(b), 18.1), communities holding GR and TK (5.2, 5.5, 7, 12.3(b), 16.1), as required by domestic ABS legislation (15.1, 17.3)	Parties concerning marine scientific research: states and competent international organizations (238)
Relationship PIC and MAT	Unresolved	PIC is linked to requirement of MAT (35)	MAT is subitem of measures for PIC (6.3(g), 17.4(g))	–
Benefit-sharing	Measures to share fairly and equitably results of research and development and benefits arising from commercial and other utilization (15.7)	Users ensure BS (16(b)(ix)); Part of MAT (45); Near-, medium, long-term benefits (47); shared fairly and equitably between all contributors (48), Mechanisms for benefit-sharing (49–50)	Measures to share fairly and equitably benefits from utilization, application, and commercialization with source/ provider country (5.1–2)	Duties to comply with certain conditions (249.1)

continued…

Table 3.1 continued

Elements	CBD	Born Guidelines	Nagoya Protocol	UNCLOS
Benefits	Participation in scientific research (15.6), Results and benefits from commercial and other utilization (15.7), Access to and transfer of (patented) technology and information (16.1, 16.3, 17.2), Joint research programmes and ventures (18.5), Participation in biotechnological research activities (19.1), Access to results and benefits from biotechnologies (19.2)	Three objectives (technology transfer, financial resources, poverty alleviation (11(g)–(h), (k)); several monetary and non-monetary benefits (16(b)(vii) and (ix), 44(d), 46, Appendix II)	Several monetary and non-monetary benefits (5.4 and Annex). Benefits directed to conservation and sustainable use (9); Collaboration in technical as well as research and development programmes, promote and encourage technology transfer to developing parties, activities in and with provider party (23)	Publication and dissemination of knowledge (244.1); Transfer of data, information, and (technological) knowledge (244.2, 268); Strengthening research and technological capacities through training, exchange of scientists, joint ventures (244.2, 268–9); Providing results, data, samples, assessments (249.1); Transfer of technology relevant to exploration, exploitation, and marine scientific research on natural resources (266)
Roles and responsibilities	Not explicitly within provisions	Explicitly for countries of origin (16(a)), users (16(b)), providers (16(c)), parties with users under jurisdiction (16(d)), and within provisions	Not explicitly within provisions	EEZ: Rights and duties for other states (58); Rights for land-locked and geographically disadvantaged states (69–70, 245); Rest not explicitly within provisions
Stakeholder involvement (including communities)	No explicit treatment	Involvement and consultation of stakeholders (14(g), 17–21, 30), Involvement facilitates monitoring (56)	Information on stakeholders held by national focal point (13.1(c)); Measures for stakeholder meetings, a help desk, involvement (21(b)–(c) and (h)); Involvement in inter-party cooperation on capacity (22.1); Support of capacity needs and priorities (22.3)	-

Elements	CBD	Bonn Guidelines	Nagoya Protocol	UNCLOS
Indigenous and local communities and traditional knowledge	Respect, preserve, maintain knowledge, innovations, and practices; promote approval and involvement of knowledge-holders; encourage benefit-sharing with holders (8(j))	An objective (protection of TK, 11(j)). Traditional uses not prevented (16(a)(iii)). Users respect TK and respond to community inquiries (16(b)(ii)–(iii)). Decisions of source state available to communities (16(a)(vi)). Source state enhance community representation (16(a)(vii)), PIC required (14(g), 31), Element of MAT (44(g))	Participate in benefits from their genetic resources (5.2); Participate in benefits from their TK (5.5); PIC or approval required (6.2, 7); Criteria and processes for obtaining PIC and involvement (6.3(f)); Considering community laws (12.1); Inform users of TK on obligations (12.2); Support development of community protocols, minimum MAT, and model contractual clauses (12.3); Not restrict customary use and exchange of GR and TK among communities (12.4); Measures for community meetings, a help desk, involvement (21(b)–(c) and (h)); Involvement in inter-party cooperation on capacity (22.1); Support of capacity needs and priorities (22.3); Measures to increase capacity of communities (22.5(j))	-
Focal points and authorities	-	National focal points informs on ABS (13), competent national authorities grant PIC (14, 28–32, 38), inform on procedures for obtaining PIC (36)	National focal point informs on PIC and MAT procedures and stakeholders (13.1(a)–(b)); CNA grants access, issues written evidence, advises on PIC and MAT (13.2); CNA provides clear and transparent written decisions, timely and cost-effectively (6.3(d))	Appropriate official channels for marine scientific research (250)

continued…

Table 3.1 continued

Elements	CBD	Bonn Guidelines	Nagoya Protocol	UNCLOS
Clearing-house mechanism	CHM to promote and facilitate technical and scientific cooperation (18.3)	An objective (CHM strengthening, 11(i)), CBD-CHM discloses focal points, national PIC procedures (13) and national access applications (16(a)(ii))	ABS-CHM as part of CBD-CHM (14.1), Receives information on national ABS measures, focal points and CNAs, issued permits and additional information (14.2–.3)	-
Awareness-raising	Promote and encourage understanding on the importance of conservation and sustainable use (13)	An objective (11(f)), Basic requirement in MAT (42(b) (i)–(ii))	Measures to raise awareness of importance of GR and TK (21)	-
Capacity	General: technical and scientific cooperation with special attention to development and strengthening national capabilities (18.2)	An objective (11(e)), Element of MAT (44(d))	Cooperation in capacity-building, -development, and strengthening of human resources (22.1); Capacity self-assessments (22.3); Capacity key areas (22.4); Measures for achieving above (22.5)	Strengthening research and technological capacities through training, exchange of scientists, joint ventures (244.2, 268–9)
Monitoring and tracking	Unrelated to GR: Monitoring of components of biological diversity and processes and activities in relation to conservation and sustainable use (7 in combination with 8–10)	CNA monitoring (14(c)), Users maintain data and documents (16(b)(vi)). National registration system (39), national monitoring (55–56), Means for verification (57–58)	Checkpoints (17.1(a)), Users and providers share information on implementation of MAT (17.1(b)), cost-effective communication (17.1(c)), 'Certificate of compliance' (17.2–.4), Element of capacity (22.5(c)), Party obligation (29)	Boarding, inspection (73.1); Reports (249.1(b)): Access to data upon request (249.1(c)); Inform of any major change (249.1(f))
Compliance	'Incentive measures' (11)	User parties take measures to support compliance (16(d)), Responsibility of collector (collecting institution) (54), Under national monitoring (55(a)), Ensured through voluntary certification schemes (57)	National measures to provide that PIC is complied with and MAT are established (GR: 15.1, TK: 16.1), Element of capacity (22.5(c))	Boarding, inspection (73.1)

Elements	CBD	Bonn Guidelines	Nagoya Protocol	UNCLOS
Violations	-	Use of sanctions (60), Parties decide on appropriate and proportionate remedies (61)	Appropriate, effective, and proportionate measures to address non-compliance (GR: 15.2, TK: 16.2), Inter-party cooperation (GR: 15.3, TK: 16.3), Opportunities to seek recourse (18.2)	Arrest, judicial proceedings (73.1); Suspension and cessation of research activities (253)
Dispute settlement	General: Settlement through negotiation, mediation, arbitration, ICJ, or conciliation (27.1–4)	Element of MAT (59)	Element of MAT (6.3(g)(i)) on applicable jurisdiction, laws, and/or alternative resolutions (18.1); Access to justice and mutual recognition and enforcement of foreign judgments (18.3)	Peaceful means (negotiation, enquiry, mediation, conciliation, arbitration, judicial settlement) (279); ITLOS, ICJ, arbitral tribunals (287.1)
Model contractual clauses and codes of conduct	Not explicitly mentioned	Standardized MTAs (42(b)(iv)) and suggested elements (Appendix I), Voluntary certification schemes (16(d)(v))	Encourage 'Model contractual clauses' on MAT (19) and 'Voluntary codes of conduct, guidelines, and best practices and/or standards' (20)	General criteria and guidelines established by international organization (251)
Distinction commercial and non-commercial research	-	An objective (facilitation of taxonomic research, (11(l), 16(b)(viii))	Promote and encourage research contributing to conservation and sustainable use, simplified access measures (8(a)), to be stated in permit (17.4(i))	EEZ and CS: research of direct significance for the exploration and exploitation of natural resources (246.5(a))
Derivatives	-	Benefits from derivatives in PIC application (36(l)), 'Derivatives' is a MAT (44(i)), Element of MTA (Appendix I B.2)	'Naturally occurring biochemical compounds resulting from genetic expression or metabolism' (2(e))	Implicitly (56.1, 77.1, 249)

continued...

Table 3.1 continued

Elements	CBD	Bonn Guidelines	Nagoya Protocol	UNCLOS
Uses of genetic resources	Research and development (15.7), Commercial and other utilization (15.7), Technologies that make use of GR (16.1), biotechnologies based upon genetic resources (19.2)	Examples: taxonomy, collection, research, breeding, commercialization (36(f), 42(e), Appendix 1 B.2)	Research and development on the genetic and/or biochemical composition of genetic resources, including biotechnology (2(c))	Exploration, exploitation, marine scientific research, but no definitions
Transfer to third parties	-	Users honour terms and conditions of acquired material (16(b)(viii)), mentioned in PIC application (36(j)), Element of MATs (44(f))	Element of MAT (6.3(g)(iii))	-
Intellectual property rights	Transfer of technology making use of GR and protected by patents and other IPR (16.3), Patents and other IPR support CBD objectives (16.5)	User states encourage disclosure of origin (16(d)(ii)), Elements of MAT (43(c)–(d))	Element of MAT on benefit sharing (6.3(g)(ii))	Protection of legitimate interests, including rights of holders and suppliers of technology (267)
Transboundary situations	-	-	Considering need for a 'Global multilateral benefit-sharing mechanism' (10), Transboundary cooperation for shared genetic resources and TK (11)	Multiple EEZs: common measures for conservation and development of stocks (63.1)
Relationship to other treaties	Affect other rights only in case of serious damage or threat to biological diversity (22.1), Consistent implementation with Law of the Sea (22.2)	In coherence with international agreements and institutions (10)	Affect other rights only in case of serious damage or threat to biological diversity (4.1), mutually supportive (4.3), subordination to specialized international ABS instruments (4.4)	No alteration as long other agreements are compatible with UNCLOS (311)

Australia (Department of the Environment, Sport and Territories, 1996, objective 2.8). To support this, Australia has developed legal frameworks that regulate the access to and utilization of genetic resources at federal, territory and state levels. At the federal level, the clarity and comprehensiveness of the legal framework merit it being regarded as model ABS legislation (Burton, 2009, p.302). However, it is not clear whether territory and state ABS legislation also meets this standard. Therefore, this and following sections will examine and compare federal, territory and state legislation on ABS, the exploration and exploitation of marine living resources and marine scientific research. At the end of this section, Table 3.2 provides a direct comparison of Australian ABS legislation at its different levels.

Federal level

Before analysing ABS legislation at the federal level, it is necessary to distinguish the geographical scope of federal, territory and state legislation. The Australian legal system operates on a federal division of powers. The power to regulate activities using natural resources in different areas is divided between the territory (Northern Territory, Australian Capital Territory), state (New South Wales, Queensland, South Australia, Tasmania, Victoria, Western Australia) and federal (Commonwealth) governments. According to Section 525(1) of the Environment Protection and Biodiversity Conservation Act 1999, areas falling under federal jurisdiction (Commonwealth areas) include: land owned by the Commonwealth; the coastal sea of Australia or an external territory; the (extended) continental shelf and the waters and airspace over the continental shelf and the waters of the EEZ, including the seabed under and the airspace above those waters. Commonwealth areas do not include areas that fall under the legislative powers of Australian states and territories. Such areas include land within the limits of states and territories as well as an adjacent belt of 'coastal waters', which extends from the baseline from which the territorial sea is measured to a line three nautical miles seaward of the baseline. Thus, the 12 nautical mile territorial sea of Australia is divided into a 3 nautical mile area under state or territory jurisdiction for the management of resources (coastal waters) and a 9 nautical mile area under Australian federal jurisdiction (Attorney-General's Department, 1980, pp.4–7).

ENVIRONMENT PROTECTION AND BIODIVERSITY CONSERVATION ACT 1999

Australia adopted the Environment Protection and Biodiversity Conservation Act 1999, which is an Act 'relating to the protection of the environment and the conservation of biodiversity, and for related purposes' (first preamble). With regard to ABS, the Act contains one section on the 'control of access to biological resources' in Commonwealth areas (Section 301). That section states that regulations may further provide for the control of access to biological resources, which include genetic resources. These regulations may contain provisions on the equitable sharing of benefits from the use of biological resources in Commonwealth areas, the facilitation of access, the right to deny access, the granting of access and conditions of access (301(1)–(2)).

With the aim of developing regulations, a public inquiry has proposed a scheme which provides for access permits and benefit sharing contracts (Voumard, 2000). Furthermore, the Commonwealth, states and territories have endorsed general principles that implement the Bonn Guidelines and that need to be applied when ABS legislation is developed or reviewed in order to establish a coherent legal framework across Australian jurisdictions (National Resource Management Ministerial Council, 2002). The Environment Protection and Biodiversity Conservation Regulations 2000 implemented this scheme and principles in their 'access to biological resources' regime within Parts 8A and 17.

ENVIRONMENT PROTECTION AND BIODIVERSITY CONSERVATION REGULATIONS 2000

The purpose of the Regulations is to provide for the control of access to biological resources in Commonwealth areas. For defining 'access to biological resources', the Regulations adopt a utilitarian approach by explaining access as the 'taking of biological resources … for research and development on any genetic resources, or biochemical compounds, comprising or contained within the biological resources'. The concept of access is further narrowed down by excluding certain uses such as the taking of biological resources by indigenous persons, access to human remains or taking genetically modified organisms or protected plant varieties or the taking of public resources. The taking of public resources involves fishing for commerce or recreation, harvesting wildflowers, taking wild animals or plants for food, collecting peat or firewood, taking oils, collecting plant reproductive material for propagation and commercial forestry (Regulation 8A.03).

Controlling access to biological resources is achieved by (8A.01):

- promoting the conservation and sustainable use of biological resources;
- ensuring an equitable sharing of benefits from the use of biological resources and ensuring that benefits accrue to Australia;
- recognizing the special knowledge of indigenous people; and
- establishing an access regime that provides certainty.

In order to have access to biological resources a person must first apply for a permit (8A.06(1)). An application form can be downloaded from the government's website (DEWHA, 2012a). An application for a permit should be submitted via email to the Genetic Resource Management Section of the Department of the Environment, Water, Heritage and the Arts. The application must include information on, e.g., the access provider (either the Commonwealth or an agency, indigenous people land holder or native title holder of land) (8A.04(1)); the biological resource to which access is sought; the amount to be taken; the use of traditional knowledge in locating particular resources or areas; the use of the biological resource and how access benefits biodiversity conservation; the conduct of access, including details of vehicles and equipment and whether the access has commercial or non-commercial aims. Fees for commercial purposes amount to AUS$50, while no fee accrues for non-commercial permits (Regulation 17.02(2)(ga)(i)–(vi) and (viii) and Schedule 11 1B.01–02).

Depending on the purposes, the Regulations require additional obligations. An applicant for a permit for (potentially) commercial purposes must additionally enter into benefit-sharing agreements with each access provider of the resource (8A.07(1)). For marine areas, the Department of the Environment, Water, Heritage and the Arts is responsible. The Department provides two model benefit-sharing agreements, which can be accessed online (DEWHA, 2012b).

A benefit-sharing agreement must provide for reasonable benefit-sharing arrangements, including the protection, recognition and valuing of indigenous people's knowledge. It must include information about the parties to the agreement; the time and frequency of entry into the relevant area; the resources (species names); the quantity to be collected and removed; the purpose of access; the means of labelling samples; the ownership of samples and transmission to third parties; any sources of indigenous knowledge used; the benefits in return for granting access and using indigenous knowledge and the benefits for biodiversity conservation in the area (8A.08) (Garforth *et al.*, 2005, p.23). The benefit-sharing agreement only takes effect if a permit for access is issued (8A.11). Where the resources occur on indigenous people's land and the access provider is the owner of land or a native titleholder for the land, the owner or titleholder must also give informed consent to a benefit-sharing agreement (8A.10(1)). In order to make sure that the access provider has given informed consent, the Minister may apply criteria. These include: whether the provider had adequate knowledge of the Regulations, engaged in reasonable negotiations with the applicant, had time to consider the application and consult with people, received legal advice and whether views of indigenous representative bodies were sought (8A.10(2)).

An applicant for an access permit for non-commercial purposes must obtain the written permission of each access provider to enter, take and remove samples of biological resources from Commonwealth areas (8A.12(1)). In addition, the applicants must submit a statutory declaration, which can also be downloaded, stating that they will prepare written reports on research results, offer taxonomic duplicates of samples, that they do not intend to use the biological resource for commercial purposes, that they will not give samples to any persons without permission of the provider and will not carry out research or development for commercial purposes unless a benefit-sharing agreement has been concluded (8A.13). The Minister then examines the applications and assesses whether certain requirements have been met (17.01(ab) and 17.03(2)(a)(aa)). In addition, the applicant must submit copies of the statutory declaration to the Minister (17.03A(6)(b)).

Where access is sought for commercial purposes, the applicant has not only to enter into benefit-sharing agreements with each provider, but they must provide copies of the agreement, and the Minister must be satisfied that the owner of the land has given informed consent to the agreement, if applicable (17.03A(6)(a)).

In addition to above requirements, it is necessary that any access benefits biodiversity conservation, is consistent with reserve management plans, is compatible with the lease of indigenous people's land and takes the precautionary principle into account (17.03A(6)(c)–(f)).

Only if these requirements are fulfilled, can the Minister issue a permit. If the Minister refuses to issue a permit, the applicant may ask for reconsideration and may take the matter to the Administrative Appeals Tribunal (Voumard, 2000, p.7 of Appendix 11). The Minister is responsible for keeping a register of information about permits (8A.18(1)), which can be accessed online.

Once the permit holder has taken samples of biological resources, they must meet additional requirements. First, the holder is responsible for keeping records for each sample. Records include: the date, place and quantity of the taking; the scientific name of the sample; the location of the sample when entered into the record; details about the disposition of the sample and a unique identifier attached to the sample or its container (8A.19(1) and (3)). In addition, a copy of the record must be sent to each access provider and the relevant department (8A.19(2)). Second, if the permit holder does not intend to keep samples, the holder must submit the samples to the access providers or to the department (8A.20).

The Regulations do not give liability provisions for cases where a person breaches obligations under benefit-sharing agreements, statutory declarations or permit requirements. However, when a person accesses biological resources without a permit, a fine of 50 penalty units is due (8A.06). Under Section 4AA of the Crimes Act 1914, one penalty unit equates to AUS\$ 110. Access without a permit would thus amount to AUS\$5,500.

The monitoring and enforcement of provisions regarding access to biological resources is unregulated within the Regulations, although some degree of monitoring can be achieved by the minister's register of permits, copies of sample records and disposal of samples to access providers and the department. However, general provisions of the Act do provide for monitoring and enforcement. For this purpose, the Minister may appoint authorized officers with monitoring powers. Officers may search premises, seize items, examine samples of substances, copy documents and use any equipment under Section 407 of the Environment Protection and Biodiversity Conservation Act 1999. Authorized officers can also enforce provisions through environmental audits, conservation orders, injunctions and civil penalties.

SEAS AND SUBMERGED LANDS ACT 1973

The exploration and exploitation of natural resources within the EEZ and on the continental shelf are regulated within the Seas and Submerged Lands Act 1973, as amended in 1994. In the EEZ, Australia has sovereign rights for the purpose of exploring, exploiting, conserving and managing all natural resources, as well as over other activities for economic exploration and exploitation (preamble). Similarly, Australia has sovereign rights for the purpose of exploring and exploiting its natural resources on its continental shelf. Australia's sovereign rights over the continental shelf are specified within the Continental Shelf (Living Natural Resources) Act 1968–1973. According to the Act, the minister may prohibit the search for or the taking of sedentary species and may grant licences authorizing such activities (Section 12–13). Additionally, an officer may enter, examine and control ships that are believed to be searching for or taking sedentary species (14). No mention is made of marine genetic resources.

FOREIGN RESEARCH VESSEL GUIDELINES 1996

However, the Foreign Research Vessel Guidelines 1996 make explicit reference to marine genetic resources. These Guidelines aim to implement the marine scientific research regime under UNCLOS. They apply to research activities in Australian waters covering the territorial sea, EEZ, fishing zone and on the continental shelf (preamble). They also apply to

> any matters relating to access to or use of genetic resources including, in particular, the collection of relatively small samples of biological material of potentially valuable compounds or attributes (including genes, foods and disease resistant plants) for conservation, [and] scientific or commercial purposes ...
>
> (Part 1)

This formulation is remarkable in three ways. First, it explicitly integrates the concept of genetic resources within a marine scientific research regime. This purports the position that the UNCLOS marine scientific research regime implicitly applies to marine genetic resources as well. Second, it indicates that activities on genetic resources are primarily related to research activities. Third, it disagrees with the general perception that marine scientific research only targets non-commercial purposes. The Guidelines rather disregard the differences between research activities for non-commercial purposes, those with direct significance for exploration and exploitation and those that are exploration and exploitation. As an effect, the same provisions of the Guidelines apply regardless of the intention of the research project.

To obtain clearance to conduct marine scientific research, researching entities must submit a request containing information on the research project six months in advance. Although it is not clear to whom such a request should be addressed, it is assumed that a request must be submitted to the Department of Foreign Affairs and Trade, because the Guidelines were released by this department.

The required information is consistent with Article 248 of UNCLOS, but the Guidelines go even further by requesting information on:

- radio, radar and sonar frequencies;
- plans for deployment and removal of equipment;
- the sharing of results, data and samples;
- radioactive and hazardous substances;
- any manipulation on the Australian continental shelf or in the EEZ, such as collecting, sampling, coring, drilling and the use of chemicals and explosives; and

- details of the 'proposed means of sharing the benefits (including financial benefits) … from the commercial development of new products derived from the discovery of natural resources in Australian maritime zones'.

The Guidelines also identify additional Australian legislation which provides obligations for foreign research vessels (Part 3). Obligations particularly relevant for activities on marine genetic resources concern research on fish and research within the Great Barrier Reef Marine Park. Foreign research vessels that intend to take, process or carry fish or wishing to search for and take sedentary organisms from the continental shelf must obtain an additional scientific research permit under the Fisheries Management Act 1991, which is issued by the Australian Fisheries Management Authority. Persons intending to undertake marine scientific research and other activities within the Great Barrier Reef Marine Park require a permit under the Great Barrier Reef Marine Park Act 1975, which is issued by the Park Authority.

REMARKS

Three main national regimes regulate access to marine genetic resources within Australian maritime Commonwealth areas: the Environment Protection and Biodiversity Conservation Act 1999, including the corresponding Regulations, which regulates access to biological resources; the Seas and Submerged Lands Act 1973, which regulates exploration and exploitation and the Foreign Research Vessel Guidelines 1996, which regulate marine scientific research.

The provisions of these regimes overlap substantially in geographical scope as well as substantive scope. However, the relationship and precedence of their provisions is not clear. The regimes do not refer to one other. The Guidelines, however, identify further Australian legislation that is relevant for activities of foreign research vessels and that apply in addition to the Guidelines. Any reference to the above-mentioned acts and regulations is, however, missing. In conclusion, all three regimes need to be implemented collectively (Lesser, 1998, p.54).

This causes a legal issue. As soon as an activity using marine genetic resources pursues commercial intentions, it is subject to both the Environment Protection and Biodiversity Conservation Act and the Seas and Submerged Lands Act. The Seas and Submerged Lands Act provides unqualified discretion for Australia to determine exploration and exploitation. This creates great legal uncertainty for users seeking access to marine genetic resources. It is recommended that Australia addresses this issue and establishes additional rules clarifying the precedence of these Acts.

Northern Territory

To date, in the Australian federal system only the Northern Territory (Biological Resources Act 2006) and Queensland (Biodiscovery Act 2004) have adopted ABS legislation for their territory and coastal waters. This does not mean that access to biological resources is left unregulated within other Australian states.

Access to marine genetic resources is often regulated under different pieces of legislation, for example, fisheries or living marine resources management acts, while benefit sharing is often left unregulated in these instruments. Nevertheless, the Australian states of Tasmania, Victoria and Western Australia have announced their intentions to develop ABS legislation (CBD, 2012).

BIOLOGICAL RESOURCES ACT 2006

The Northern Territory has adopted the Biological Resources Act 2006, which is 'an act to provide for and regulate bioprospecting in the Territory and for related purposes' (preamble 1). 'Bioprospecting' is similarly defined as in federal legislation as the 'taking of samples of biological resources … for research in relation to any genetic resources, or biochemical compounds, comprising or contained in the biological resources'. In addition, certain activities are excluded, such as the taking of biological resources by indigenous people, dealing with material of human origin, taking samples for a purpose other than biodiscovery, taking genetically modified organisms or protected plant varieties, taking aquatic life under a permit granted by the Northern Territory Fisheries Act 1988 (excluding special permits) and the taking of public resources, as defined under federal law (Section 5(1)).

The main object of the Act is to facilitate bioprospecting by (3):

* promoting the conservation and sustainable use of biological resources;
* establishing an access regime designed to give certainty;
* establishing a contractual framework for benefit-sharing agreements between persons engaged in bioprospecting (bioprospectors) and resources access providers;
* recognizing special knowledge held by indigenous people; and
* seeking to ensure that benefits accrue to the Northern Territory.

Before bioprospecting can take place, the person intending to engage in bioprospecting must submit an application to the appropriate issuing authority to obtain a permit to take biological resources (11). The authority considers whether the proposed activity comprises bioprospecting and then refers the application to the Chief Executive Officer (CEO) of the agency administering the Act (12). If the CEO considers the proposed activity to comprise bioprospecting then the applicant must enter into a benefit-sharing agreement with each resource access provider (6, 16, 19).

A benefit-sharing agreement must provide for reasonable benefit-sharing arrangements. It must include details about: the parties of the agreement; the time and frequency of entry into the area; the resources (species names or lowest level of taxon, if known); the quantity of the resource to be removed; the purpose of access; the disposition of ownership in the samples, including transmission to third parties; the use of traditional knowledge and the benefits to be given in return and the benefits that the provider will receive in return for granting access (29.(1)).

A benefit-sharing agreement is only valid if the provider has given prior informed consent to the terms of the agreement (27(3)). The CEO must be satisfied the provider has given prior informed consent, and must therefore consider whether the provider has adequate knowledge of the Act, has engaged in reasonable negotiations with the applicant, has had adequate time for consultation and has received independent legal advice on the implementation of the Act (28). When the CEO is satisfied that a suitable benefit-sharing agreement is in place, the authority may issue the applicant with a permit to take the specified biological resources (20, 21), and the CEO maintains a public register of permits issued, samples taken, benefit-sharing agreements and certificates of provenance (23(1), 33). A certificate of provenance is a document issued by the CEO, confirming that the biological resource was taken under a permit scheme, with the consent of the provider, and that a benefit-sharing agreement was in place (36(2)). Once a permit has been issued, the benefit-sharing agreement also comes into effect (32).

After sampling biological resources, the bioprospector must report to the authority on the date, location, species and quantity of the sample (24–25).

Finally, the Act provides for various liability provisions. If a bioprospector takes biological samples without a permit, gives false information or breaches permit conditions or a benefit-sharing agreement, a maximum penalty of 500 units is due (38–41). The monetary amount of a penalty expressed in penalty units is obtained by multiplying the units by the prescribed value of AUS$130, as laid down in section 3 and 4(1) Northern Territory of Australia Penalty Units Act 2009. Thus 500 penalty units equal to US$68,000. If a bioprospector does not keep a record of information for each sample (42(1)(a)–(g)), provide copies of records to providers, the relevant authority and the CEO within reasonable time or incorrectly disposes of samples, then a maximum penalty of 100 penalty units is due (US$13,500) (42–43).

FISHERIES ACT 1988

The Northern Territory Fisheries Act 1988 is also relevant to access to marine genetic resources. This Act provides regulations for the issuing of a special permit. The special permit enables the holder to take fish or aquatic life for the 'purposes of education, research or the carrying out of trials and experiments with fishing vessels or fishing gear or any other apparatus or technique, which is capable of being used in connection with the taking of fish or aquatic life' (17(1)(a) and (c)).

Queensland Biodiscovery Act 2004

Queensland has adopted the Biodiscovery Act 2004, which is 'an act about taking and using State native biological resources for biodiscovery, and for other purposes'. The Schedule of the Act defines 'biodiscovery' in two ways: it is either the analysis of molecular, biochemical or genetic information about native biological material for the purpose of commercializing the material, or it is the commercialization of a product from biodiscovery research. 'Commercialization' means 'using the material in any way of gain'. 'Gain' is not defined. While 'commercialization' indicates that the Act applies only to activities with economic gains, '*any* way of gain' (emphasis added) can be assumed to also include non-economic gains, such as knowledge, as well.

The main purposes of the Act are to facilitate access to native biological resources for biodiscovery, to encourage the development of value-added biodiscovery, to ensure that the State obtains a fair and equitable share in the benefits of biodiscovery and to ensure that biodiscovery enhances knowledge of the State's biological diversity (3(1)). These purposes are achieved by establishing (3(2)):

- a regulatory framework for taking and using native biological resources for biodiscovery;
- a contractual framework for benefit-sharing agreements;
- a compliance code and collection protocols; and by
- monitoring and enforcement of compliance.

For the taking of native biological resources, the Act introduces the concept of 'collection authorities', which are permits to take and use native biological material for biodiscovery (10). A person must submit an application for an authority to the chief executive of the department that administers the Queensland Nature Conservation Act 1992 in the approved form and accompanied by a biodiscovery plan (11(1)–(2)). The approved form of the application includes the applicant's name and business, the geographical area for which access is sought, a detailed description and scientific classification of the material and the proposed period for which the authority is sought (maximum 3 years) (12, 16(2)). The biodiscovery plan is a description of proposed biodiscovery activities and provides information on commercialization activities, a timetable for activities, activities that are conducted outside the state, other parties engaged by the applicant and potential benefits provided under a benefit-sharing agreement (37(a)–(e)).

After considering the application, the chief executive may either refuse or grant the application (14(1)). In case the chief executive decides to refuse an application, the applicant can appeal against the decision and submit the application to the minister for internal review (95–98). The chief executive can only grant the application if the proposed taking and use is for biodiscovery (14(2), 17(2)) and it conforms to the compliance code (44) and applicable collection protocols (45(1)–(2)). Granting the application and issuing an authority (18, 20–23, 26–28) does not empower the holder of the authority to take native biological material. The holder also needs to enter into a benefit-sharing agreement (17(1)).

The Minister administering the Queensland Gene Technology Act 2000 enters into benefit-sharing agreements with a biodiscovery entity (Schedule, 33(1), 36, 38–41), which is, for example, a holder of a collection authority. In combination with an authority, the agreement gives the right to take material and to use it for biodiscovery. The agreement obliges the entity to provide benefits of any biodiscovery to the state (17, 33(1)(a)–(b)). It contains information on the date, term, (monetary) benefits to be provided to the state, the number of the collection authority and the place of business of the biodiscovery entity (34(2)(a)–(f) and (h)). The chief executive of the authority administering the Queensland Gene Technology Act 2000 keeps a register of benefit-sharing agreements, and the biodiscovery entity must keep records of results of biodiscovery carried out under the agreement for 30 years (42–43).

After taking native biological material, the holder of a collection authority (the biodiscovery entity) must comply with various obligations. First, the holder is obliged to label the material (29(1)). The label must contain information on the number of the authority, the date, the material's scientific classification and the geographic location from which the material was taken (29(2)(a)–(d)). Second, any samples of material must be sufficiently identified so that their source can be tracked (29(3)). Third, the holder must give a sample of collected material to the state (30(1) and (3)). Fourth, the holder must submit a semi-annual material disposal report to the chief executive. The report contains information about all native biological material taken and given to another party (32(1)(a)–(b)).

The Act further provides for penalties concerning offences against collection authorities, biodiscovery plans and benefit-sharing agreements. Should a person collect native biological material for biodiscovery without authority, a maximum penalty of 2000 units is due (50(1)(b)). According to Section 5(1)(c) Queensland Penalties and Sentences Act 1992, one penalty unit equals US$105 (maximum US$210,000). If that material originates from endangered, rare, vulnerable or protected animals, the maximum penalty can rise to 3,000 penalty units or two years' imprisonment (50(1)(a)). If a person contravenes a condition, gives false or misleading information or supplies false or misleading documents concerning an authority, a biodiscovery plan or a benefit-sharing agreement, a penalty of 100 penalty units is due (51–53, 55–57). The highest penalty applies for using material without a benefit-sharing agreement, and it amounts to either 5,000 penalty units or the full commercial value of commercialization, whichever is greater (54(1)). The Act contains many additional penalties within the various obligations laid down by it.

For monitoring the implementation of the Act and enforcement in cases of non-compliance, the chief executives may appoint inspectors (61(1)). The inspector has the power to seize items that are evidence of an offence against the Act (78(2) and (4)). The inspector may enter buildings (68(1), 69, 70–73), stop vehicles, boats and aircraft in order to search and inspect them, may copy any documents and use any equipment necessary in the execution of their powers (74(1)–(2), 75(3)(a)–(d)).

Finally, the Act contains provisions for legal proceedings (108–116).

Other Queensland acts that may also concern access to marine genetic resources do not apply in addition to the Biodiscovery Act 2004. Where a collection authority for taking native biological material is issued under the Act, a person is not required to obtain a licence, permit or other authority for same material under a different act (7(2)(a)). A provision with similar effect can be found under the Queensland Fisheries Act 1994. This Act does explicitly not apply to 'the taking and keeping of fish [including crustaceans, molluscs, sponges and corals] for which a collection authority has already been issued under the Biodiscovery Act 2004' (5(1)–(2), 11(2)(e)).

European ABS legislation and policy measures

Like Australia, European states are primarily user states. User states are typically perceived to be poor at implementing international ABS provisions, especially user measures (Young, 2008, p.117). However, an increasing amount of ABS-relevant policy and legislative measures has emerged in Europe on the supranational as well as on the national level, despite the fact that European states have upheld the notion of free access to genetic resources for a long time (Kamau *et al.*, 2010, p.260). It is therefore particularly interesting to examine whether European states increasingly regulate access to their genetic resources only or also the behaviour of users within their jurisdiction. The assessment includes an overview of all measures currently available within the European Union (EU), selected member states (Belgium, Bulgaria, Germany, Italy, Malta and Portugal) and some associated states (Greenland and Norway).

European Union

The EU has not yet introduced comprehensive legislation governing ABS. However, various policy and legislative measures do address the implementation of international ABS provisions.

On the policy side, there are four major documents. The Biodiversity Strategy of the European Community (EC, 1998) recommends the promotion of cooperation between countries to guarantee access to natural resources, technology transfer and scientific and technical cooperation. The Biodiversity Action Plan stated that the EC will support national capacity building and develop laws for equitable benefit sharing (EC, 2001, para. 39). In 2003, the European Commission produced a communication document on the implementation of the Bonn Guidelines (EC, 2003). In order to encourage users to fulfil their obligations under the CBD, the Commission considered: the creation of a European information network of ABS focal points and an EC Biodiversity Clearing House Mechanism (EC-CHM, 2012); the disclosure or certificates of origin; arbitration of infringements and the development of voluntary certification schemes (EC, 2003, pp.13–23). Finally, the 2008 EU Biodiversity Action Plan called for a reduction in the impact of international trade on global biodiversity by, among others, promoting full implementation of the Bonn Guidelines and other similar international agreements (European Commission, 2008, p.21).

On the legislative side, only the 1998 EC Directive on the Legal Protection of Biotechnological Inventions (98/44/EC) states within its recitals that patent applications should include information on the geographical origin of material used as the basis for inventions (Recital 27). However, recitals do not constitute legally binding obligations for member states (EC, 2003, p.10).

The Directive is established on the exclusive ability of the EU to institute the rules necessary for the functioning of the internal market as laid down by Articles 3 and 114 of the Treaty on the Functioning of the European Union. At the same time, the EU possesses explicit powers for intellectual property under this Treaty, which also aims at the functioning of the internal market and the

Table 3.2 Categorization of ABS-relevant legislation of Australia (Environment Protection and Biodiversity Conservation Act 1999 and Regulations 2000), the Northern Territory (Biological Resources Act 2006) and Queensland (Biodiscovery Act 2004). Relevant regulations and sections in parentheses. CE: Chief executive, CEO: Chief executive officer, DBE: Dept. of Business and Employment, DoR: Dept. of Resources, DSDI: Department administering the Queensland Gene Technology Act 2000, EPA: Department administering the Northern Territory Biological Resources Act 2006, PUs: Penalty units

Elements	Australia, federal level	Northern Territory	Queensland
Major aim	An Act relating to the protection of the environment and the conservation of biodiversity.	An Act to provide for and regulate bioprospecting in the Territory.	An Act about taking and using State native biological resources for biodiscovery.
Purposes/objects	Control access to biological resources (8A.01 of the Regulations).	Facilitate bioprospecting (3(1)).	a) Facilitate access, b) encourage biodiscovery development, c) ensure benefit sharing, d) enhance knowledge of biodiversity (3(1)).
Means to achieve purposes	a) Conservation and sustainable use, b) access regime, c) equitable sharing of benefits, d) traditional knowledge, e) benefits accrue to Australia (8A.01).	a) Conservation and sustainable use, b) access regime, c) equitable sharing of benefits, d) traditional knowledge, e) benefits accrue to Territory (3(2)).	a) Framework for taking and using, b) framework for benefit sharing agreements, c) compliance code, d) monitoring and enforcement (3(2)).
Regulated activity	Access to biological resources (8A.03).	Bioprospecting (3(2) and 5).	Taking and using native biological material for biodiscovery (3(2)).
Application for permit	'Permit' (8A.06, Part 17).	'Permit' (11–5, 20–3).	'Collection authority' (10–32).
Prior informed consent	'Informed consent' by access provider (8A.04, 8A.10).	'Informed consent' by resource access provider (6, 28).	-
Mutually agreed terms	Benefit-sharing agreement (8A.07, 8A.08); Permission and statutory declaration (8A.12–13).	Benefit-sharing agreement (27, 29).	Benefit-sharing agreement (17.1, 33–5).
Relevant authorities	Dept. of the Environment, Water, Heritage and the Arts (8A.04–07, .10, .12, .15–.18, .20, 17.03(2)(a)).	'Permit issuing authority' (DoR and Dept. of natural resources, environment, the art and sports) (4, 11–3, 20–5, 44); 'agency administering this Act' (DBE) (13–23, 25–6, 28, 30–1, 33–7, 44–6).	'EPA' for collection authorities (11, 13–5, 17, 19, 20–3, 26–8, 44–7, 49, 52–3, 61, 95–8, 111, 122, schedule); 'DSDI' for benefit-sharing agreements (32–3, 36, 38–42, 56–8, 61, 99–102, 111, 122, schedule).

Elements	Australia, federal level	Northern Territory	Queensland
Compliance provisions	-	Undefined 'standards of operation' (17(2)(a)).	'Compliance code' (44, 46–9); 'Collection protocol' (45–9).
Obligations after access	Keep and submit sample records (8A.19), correct sample disposal (8A.20).	Sample report after access (24), keep sample records (42), correct sample disposal (43).	Return authority upon cancellation (22), label material and identify sample (29), submit samples to state (30), material disposal report (32), record biodiscovery results (43).
Registers	Minister's register of permits (8A.18).	CEO's register of permits (23, 33); benefit-sharing agreements (31, 33); samples (33); certificates of provenance (33).	EPA CE's register of collection authorities (27), DSDI CE's register of benefit-sharing agreements (42).
Ownership/exclusive rights	Ownership of sample subject to benefit-sharing agreement (8A.08).	-	No exclusive rights through authority or benefit-sharing agreement (44).
Appeal of decisions	Administrative Appeals Tribunal.	N/a, decline of applications should only be explained (18(2), 21(4)).	'Internal reviews' of decisions (14, 40, 95–102).
Monitoring and tracking	Register (8A.18; copies of agreements and declarations to the Minister (17.03A(6)), records and samples to access providers and Dept. (8A.19); monitoring by authorized officers (407 of the Act).	Report after access (24), registers (33), provision of sample records (42).	Registers (27, 42); labelling of material (29), semi-annual material disposal report (32); 'Monitoring and enforcement' through inspectors (61–94).
Offences	Access without permit, 50 PUs (8A.06).	'Offences', 100–500 PUs (38–43).	'Offences', 20–5000 PUs (22, 29–30, 32, 43, 50–60, 67, 74, 76–7, 81, 91–4, 117).
Enforcement	'Enforcement' by authorized officers, audits, orders, injunctions, and civil penalties (Part 17 of the Act).	-	'Inspectors' (61–94).
Legal proceedings	'Civil penalties' (481–486D of the Act).	-	'Legal proceedings' (108–15).

continued…

Table 3.2 continued

Elements	Australia, federal level	Northern Territory	Queensland
Distinction commercial and non-commercial	Access for 'commercial purpose' (8A.07–.11) and 'non-commercial purpose' (8A.12–.14).	No distinction	No distinction, but 'biodiscovery' aims at commercialization (schedule).
Relationship to other acts	Yes, but irrelevant for ABS (9 of the Act).	n/a	'Fisheries Act 1994' (7).
Derivatives	'taking of biological resources for research and development … on biochemical compounds' (8A.03).	'taking … of biological resources…for research in relation to … biochemical compounds' (5)	'analysis of molecular, biochemical or genetic information' (schedule).
Coverage of marine genetic resources	'Commonwealth areas' include territorial sea, EEZ, and continental shelf (8A.02, 525 of the Act).	'throughout the territory (including the water and seabed …).' (9, 5(2)(a), 6(1)(e)).	'Queensland waters' (10, 12, 27, 46, 50, 54, 124, 135, schedule).

uniform protection of intellectual property (118). The control of the EU in this area seems reasonable for a number of reasons. First, it supports the EU's efforts to harmonize patent law across member states, e.g., the European patent. Second, disclosure requirements are a controversial topic among member states and could be better realized at an EU-level. However, additional measures are necessary to ensure effective benefit sharing, such as legally binding disclosure requirements, the integration of certificates of origin or other measures that regulate users and aim at functioning of the internal market.

With regard to regulating access to genetic resources, the EU has not adopted any ABS measures, and regulation remains the responsibility of national access laws. This might be the best approach to ensure relevance to local realities and needs (EC, 2003, p.13). However, the actions of the EU can be justified either on grounds of removing obstacles to the internal market or on the competence of the EU to preserve, protect and improve the quality of the environment as found under Articles 191 and 192 of the Treaty.

Belgium

No specialized legislation on the access to (marine) genetic resources could be identified. In general, Belgium has proclaimed an EEZ and sovereign rights and jurisdiction within Article 2 of the Act Concerning the Exclusive Economic Zone of Belgium 1999. The Act also prescribes that foreign marine scientific research requires the consent of the Minister responsible for foreign affairs, without specifying further steps in this regard (40).

However, Belgium has adopted user measures. According to the Belgian Patent Act 1984, as amended in 2005, anyone who wants to obtain a patent must submit an application (13). With regard to the utilization of genetic resources, the patent application must contain a reference to the geographical origin of the plant or animal biological material, if known, from which the invention has been developed (15.1(6)). Under the law, 'biological material' refers to material containing genetic information and capable of reproducing itself or being reproduced in a biological system (1.1).

Bulgaria

The Bulgarian Biological Diversity Act 2002 regulates conservation and the sustainable use of biological diversity (1(1)). According to the Act, the state of Bulgaria owns the genetic resources of natural flora and fauna (66(1)). Access to its genetic resources requires advance agreement on the terms of benefit sharing, including citation of the material, provision of results, recovery of parts and derivatives of the resources and participation in joint scientific studies (66(3) and 66(6)). In addition, the provision of material to third parties requires the consent of the owner (66(5)). Finally, any access to genetic resources intended for non-commercial purposes is supported by offering the chance of gratuitous provision (66(4)).

Bulgarian law contains provisions on marine scientific research within the Act Governing the Ocean Space of the People's Republic of Bulgaria 1987.

This Act repeats the provisions laid down by UNCLOS within provisions on the authorization of marine scientific research, the obligations of foreign nationals when conducting research and the suspension or cessation of research (55–57).

Germany

Like Belgium, Germany has not adopted specialized access legislation for (marine) genetic resources but has instead adopted first user measures. The German Patent Act 1936, as amended in 2009, requires any inventions that contain or use biological material of plant or animal origin to include a declaration of the geographical origin, if known (34a). The examination of the application and validity of the patent's rights remains unaffected by the declaration.

Given the absence of additional legislation, it is noteworthy that two policy documents provide further guidance on ABS. The Federal Ministry for the Environment, Nature Conservation and Nuclear Safety has published the National Strategy on Biological Diversity. The Strategy mentions the aspiration to continue national negotiations on implementing the Bonn Guidelines with special emphasis on compliance with prior informed consent and mutually agreed terms (Federal Ministry for the Environment, 2007, p.58). The German Research Foundation has released the Guidelines for Funding Proposals Concerning Research Projects within the Scope of the CBD. The Guidelines provide information enabling scientists to comply with international and national ABS provisions when conducting research projects (DFG, 2010, p.1).

Italy

The only relevant measure available in Italian law concerns the implementation of the disclosure requirement from the above-mentioned EC Directive. Under Article 5.2 of Law No. 78 from 2006, the origin of any biological material of animal or plant origin and which is the basis of inventions should be declared at the time of patent application, mentioning the country of origin.

Malta

Malta adopted its Flora, Fauna and Natural Habitats Protection Regulations 2006. Under the Regulations, access to genetic resources is subject to the prior informed consent of the competent authority (36(1), 43–48). Access should be on mutually agreed terms and should guarantee fair and equitable sharing of benefits from research, development and commercial utilization (36(2)–(3)) Where a person contravenes these regulations, they are liable to a fine of between Lm 1,000 and 20,000 (49(3)) (Lm 1,000 are approximately US$3000).

Portugal

In 2002, Portugal adopted the Decree-Law no. 118/2002, which establishes a legal regime for the registration, conservation, legal safeguarding and transfer of autochthonous (native or indigenous) plant material (1–2). Under the Decree, any sub-national provider may apply to the Director General of Crop Protection for a chargeable registration of plant material, to be filed at the National Center for the Registration of Protected Varieties (4(1), 4(5)–(6), 9(1), 16). Sub-national providers can be public, private, individual or corporate entities, provided that they represent the interests of the geographical area, where the local variety of plant material is most widely found. Granted registrations confer various rights and duties upon the provider. The provider must apply before access to and use of plant material take place, is then entitled to receive a share of the benefits derived from use of the material and is responsible for in-situ maintenance of the material (4(4), 7(1)–(2), 7(4), 10(1)–(2)).

For the access and use of plant material, the prior authorization of the Technical Council of the Ministry of Agriculture, Rural Development and Fisheries on Agrarian Genetic Resources, Fisheries and Agriculture needs to be obtained (7(1)–(2)).

The protection of traditional knowledge (3(1)) plays a prominent role within the Decree. As with the registration of plant material, holders of traditional knowledge can also register their knowledge (3(2)). Registration affords the owner the right to: a) protect their knowledge against reproduction or commercial or industrial use and b) assign, transfer and license rights in traditional knowledge (3(2), 4(4)).

Finally, the Decree contains compliance measures through the imposition of fines (US$130–38,000) (13) or sanctions (loss of property, prohibition from practising a profession, removal of various rights or suspension of authorizations, licences and permits) (14).

With regard to marine scientific research, Portugal adopted Decree-Law No. 2/81. The Decree prescribes that marine scientific research activities by national private entities, other states, public entities, private foreigners or international organizations within the EEZ require the authorization of the Portuguese state (1). Researching entities must submit a request for authorization to the Ministry of Foreign Affairs six months before the commencement of the proposed activity (3). The request must contain the information as required by UNCLOS. Concession of authorization obliges the holder to supply the Portuguese state with any processed data, preliminary reports, final results and the conclusion of the completed work (5). Where the holder of an authorization violates the conditions of their authorization, the Portuguese state may demand the immediate suspension or cessation of research activities (6).

Greenland (Denmark)

Greenland enacted the Greenland Home Rule Parliament Act no. 20 on Commercial and Research-Related Use of Biological Resources 2006. The Act aims mainly to ensure the exploration of biological resources (3) is conducted in accordance with the CBD, utilizing research results to create commercial value and ensuring that Greenland obtains a fair share of values created from biological resources (1). Values arising from non-technical uses are excluded, such as hunting, fishing, agriculture and collection for decoration and consumption. The Act therefore introduces the concept of survey and commercial licences.

A survey licence is required before an entity can engage in any acquisition, collection or survey of biological resources in connection with research or subsequent commercial utilization (6(1)). Applications, as annexed to the Act, must be submitted to the Greenland Ministry of Infrastructure and Environment, which also grants the licences (Greenland Government, 2012). The Ministry prescribes the terms of a survey licence related to the impact of collection on the environment, area of collection, methods, reporting obligations and transfer of rights to the Greenland government (6(4)). The Ministry can also enter into additional private-law agreements with the applicant (6(5)). Where publications arise from survey results, the Ministry must be informed and receive copies of the publication (8(1)–(2)).

If commercial utilization is the aim, a commercial licence from the Greenland Ministry of Industry, Labour, Vocational Education and Training is required in addition to the survey licence (10(1)). The Ministry may also enter into private-law agreements to implement more detailed terms of utilization (10(3)). These could pertain to the scope and duration of the project, reporting obligations, termination of utilization and payment (11(1)). Payments for commercial licences are calculated on the basis of duration and the expected financial return on the licence (11(2)).

The Act further obliges holders of survey or commercial licences to submit annual reports and to inform the government should the holder pursue patent protection for an invention (9, 12).

Finally, the Act provides for fines in cases of: acquisition, collection and surveys conducted without a survey licence; transfer and communication of survey licences to third parties; commercial utilization without a commercial licence and violation of publication and reporting obligations (16(1)). These fines also apply to enterprises fully or partially owned by the state (16(2)).

It is not clear whether this Act applies only to terrestrial genetic resources or also to marine resources. The only clear indication can be found in the Act's definition of the collection of biological resources, which includes the collection of sea ice. Given the absence of a clear geographical definition, it is worth referring to the Exclusive Economic Zone Act 1996, which established: a) sovereign rights over the economic exploration and exploitation of natural resources and b) jurisdiction over marine scientific research in Denmark's EEZ (3). Special legislation in this regard has not been identified.

Norway

Norway is the country of origin of various commercially successful drugs. However, no benefits have accrued to Norway because neither the CBD nor any national ABS legislation was in force at that time (Svarstad *et al.*, 2000). Today, there are four laws which contain relevant provisions on the collection and utilization of marine genetic material, which can be traced back to the Nordic Ministerial Declaration on Access and Rights to Genetic Resources.

The Act Relating to the Management of Biological, Geological and Landscape Diversity 2009 aims at protecting diversity and ecological processes through conservation and sustainable use (1). The Act applies to both land territory and the Norwegian territorial sea (2). Within Chapter 7, the Act regulates access to genetic material. According to the Act, genetic material (3(f)) is a common resource belonging to the Norwegian people and managed by the state (57). In order to obtain a permit for the collection or utilization of genetic material, it is necessary to apply to the Ministry of the Environment (58). The monarch, being the highest authority, may make regulations about the type of information these applications must contain, such as the use of indigenous knowledge, and conditions on any benefits to ensure that they accrue to the state (58). Furthermore, genetic material can also be held by public collections, which are publicly accessible (59).

It is remarkable that the Act adopts various measures regulating the activities of Norwegian nationals involved in the import of genetic material obtained abroad (60). The import of genetic material from states requiring consent for collection and export can only take place with such consent. The person in control of the genetic material is bound by the conditions under which consent was originally granted, and provider states may enforce the conditions through legal actions. In addition, the utilization of genetic material must be accompanied by stating the provider country and whether consent has been obtained, if it is required. In cases where the provider country and the country of origin are not the same, information on the country of origin and whether consent for collection was granted by them also needs to be stated. Although these regulations constitute pioneering work towards user regulation, there are still several gaps and weaknesses. First, it is still difficult to identify the illegitimate use of genetic materials. Provider states often lack the capacity to monitor utilization and must rely on user compliance; however, compliance measures such as fines, sanctions or incentives are lacking in Norway, as are institutions to control utilization, such as checkpoints. The competent national authority could serve as a main checkpoint but it lacks the mandate to do so. Second, even if illegitimate uses are discovered, there are no clear rules on legal proceedings or on how and where provider states can enforce their rights. In addition, there seems to be no support from Norway in such cases, so that provider states are left on their own to penetrate the Norwegian law and to carry the costs of enforcing their rights. Again, extending the mandate of the competent national authority to provide at least legal advice might improve this situation. However, while these deficiencies are not resolved, these user measures remain largely ineffective.

The 1967 Norway Act No. 9 on Patents contains a provision with similar effects. Under this Act, patent applications for inventions using biological material must include information on the providing country, country of origin (where it differs from the providing country) and whether prior consent has been obtained (8). A statement of prior consent is only necessary if the providing country or country of origin require consent for access to biological material. If the country of origin is not known, the application must state this as well.

The 2008 Norway Act No. 37 Relating to the Management of Wild Living Resources contains some provisions on marine bioprospecting. The Act's purpose is to ensure sustainable and economically profitable management of wild living marine resources and the genetic material derived from them (1). Its geographical scope does not only cover internal waters and the territorial sea, but also the continental shelf and the Economic Zone of Norway (4). Within these areas, harvesting and investigations in connection with marine bioprospecting require a permit from the Ministry of Fisheries and Coastal Affairs (9). The monarch may adopt regulations prescribing further information applications must include conditions on marine bioprospecting. With regard to benefit sharing, the conditions of a permit may prescribe that any benefits from the use of marine genetic material accrue to Norway (10). In addition, a permit can forbid the selling of any genetic material and the communication of the results of bioprospecting activities to third parties without the consent of Norway or an agreed payment.

In 2001, Norway adopted the Regulations Relating to Foreign Marine Scientific Research, which follow UNCLOS provisions closely. The main purpose of the Regulations is to promote marine scientific research and increase scientific knowledge of the marine environment (1). Foreign marine scientific research within the internal waters, territorial sea, economic zone and on the continental shelf requires an application to be submitted to the Norwegian Directorate of Fisheries (3, 6, 8), which is annexed to the Regulations. Consent from the Directorate can be linked to compliance with certain conditions, such as the participation of Norwegian scientists, the submission of reports and final results and access to data and samples (11). Furthermore, the researching entity is under the obligation to inform the Directorate of any changes, to comply with inspection and to suspend or cease the research under certain circumstances (14–15, 21–22).

It is not clear how these Regulations relate to the bioprospecting regime within the Marine Resources Act 2008. The Regulations refer to other acts that override the Regulations in cases where the studies are of direct significance for the exploration and exploitation of natural resources, but a reference to the Marine Resources Act 2008 is missing. Future regulations on marine bioprospecting or amendments of the Regulations should clarify this relationship.

Remarks

The analysis of these selected examples of ABS legislation allows for a number of conclusions.

First, states with users under their jurisdiction employ considerably different ABS-measures. Most employ either provider measures, which regulate access or user measures, which regulate users. Only Norway followed a dualistic approach by combining provisions that regulate access to genetic resources as well as the behaviour of users within its jurisdiction. Because any state globally can be both a provider of genetic resources and also accommodate potential users, any legislation focusing on either aspect is incomplete and fails to ensure a fully functioning ABS regime.

Second, provider measures, where adopted, also vary significantly. While some countries, such as Australia, have adopted very extensive and comprehensive legislation, most lay down only very basic principles leaving many issues unresolved. There are lacunae in the definitions of key terms and concepts, the geographical scope and coverage of the marine environment, application procedures, the participation of stakeholders, the treatment of traditional knowledge, the allocation of benefits, the distinction between research for commercial and non-commercial purposes, compliance, obligations after access, monitoring and tracking, enforcement, the appealing of decisions and the relationship with other acts when multiple acts regulate same activity.

Third, there is a recognizable yet very weak trend towards the adoption of user measures. The most basic measure is the declaration of origin within patents. The fact that countries have already adopted this measure is remarkable because currently neither international nor European supranational law oblige states to do so. Nevertheless, the benefits that such a declaration entails for a provider state can be disputed, and more extensive user measures are required.

Fourth, the relationship between ABS legislation and other legislation applicable to the collection and utilization of marine genetic resources, e.g., marine scientific research, is rarely resolved. In cases where a country has adopted domestic legislation on ABS covering marine genetic resources and marine scientific research, access seekers must comply with both (Arico and Salpin, 2005, p.277). This creates a very cumbersome situation for access seekers who struggle to know all the relevant acts with which to conform. The need to obtain a range of permits yet always being at risk of not having applied for all relevant permits entails legal uncertainty and deters users from pursuing access within a certain country.

Fifth, several of the analysed laws contain provisions which could be used as model elements for ABS legislation. For regulating access, particular reference is being made to Australian law at federal, state and territory levels, although these have also been criticized for their strictness deterring research and development (ten Kate and Laird, 2000b, p.301). On the positive side, these laws create transparency and legal certainty by establishing detailed procedures around access and benefits to be shared. The rights and responsibilities for both access seekers and providers are clearly stated. The participation of stakeholders and the recognition of traditional knowledge (except for in Queensland) increase public acceptance. Monitoring, reporting and labelling obligations for users improve the traceability of samples. A clear distinction between commercial and non-commercial activities (currently only at the federal level) fosters the development of basic research. The opportunity to appeal against decisions and the maintenance of public registers, through which prior informed consent and mutually agreed terms can easily be verified, are both attractive for potential users. And clarifying the relationship with other acts simplifies access procedures.

For regulating user behaviour, Norway constitutes a good example for the way forward. Norwegian legislation not only prescribes a declaration of origin, but it also requires users to comply with the laws of provider states and grant provider states the right to engage in legal actions. The adoption of such measures without international obligations, e.g., under the Trade Related Aspects of Intellectual

Property Rights (TRIPS) Agreement of the World Trade Organization (WTO), disarms to a certain extent the criticism that such measures would amount to a 'legal and administrative nightmare' (Jeffery, 2002, p.773). However, implementation will be difficult. It lacks measures on monitoring, compliance and enforcement, and it seems that provider states are solely responsible for discovering illegitimate uses and enforcing their rights.

Finally, it is extremely necessary that states with users under their jurisdiction continue to adopt measures which regulate user behaviour. The Bonn Guidelines and the Nagoya Protocol provide the best available orientation for such measures.

Selected case studies

In order to illustrate how ABS regulations influence agreements between users and providers of genetic resources, two case studies will be looked at in detail. They deal with the sampling of marine genetic resources from waters under Australian jurisdiction.

Biological resources access agreement between the Commonwealth of Australia and the J. Craig Venter Institute

From 2004 to 2006, the J. Craig Venter Institute conducted the second Sorcerer II Global Ocean Sampling Expedition (Rimmer, 2009, pp.171–183; Australian Government, 2012; CAMERA, 2012; JCVI, 2012). The purpose of the Expedition was to advance scientific knowledge of microbial biodiversity and increasing humankind's basic understanding of ocean biology, including the interplay between micro-organisms with and their effects on environmental processes. To achieve this, the Sorcerer II circumnavigated the globe and took water samples from 23 different countries. The marine micro-organisms contained within the water samples were analysed for their genomic sequence in order to find new species and genes. Data from the first cruise resulted in the discovery of 1,800 species and 1.2 million genes.

In 2004, the Expedition intended to collect samples in maritime areas which fell under the jurisdiction of the Commonwealth of Australia. The Institute therefore submitted an application for consent to conduct marine scientific research. As a result, Australia and the Institute negotiated and adopted the Biological Resources Access Agreement Between the Commonwealth of Australia and the Craig J. Venter Institute, which provides the terms under which biological resources can be taken and used.

Under the Agreement, Australia granted a non-exclusive licence to the Institute to research collected material (seawater, sediments and any biological resources contained within it) (Schedule section 3) and results (all information and tangible objects arising from the use of the material and any derivatives) (Section 1.1(k)) for the purposes of: inventorying micro-organisms; understanding overall species diversity; discovering new bacterial and viral species; evaluating ecological roles of microbes and establishing a freely shared global environmental genomics database (Sections 4.1, 5.3, Schedule section 6). Without the prior written permission of Australia, the Institute cannot use either materials or results for any other purposes nor can it provide any materials or results to third parties. Furthermore, the research is purely non-commercial (Section 4.6). However, if the Institute decides to commercialize its results or intellectual property arising from the use of materials, it must first enter into an appropriate agreement with Australia. Similarly, Australia must be notified as soon as possible about any inquiries for commercial purposes received from third parties (5.7). All property rights, including intellectual protection, relating to materials and results are vested in Australia and not in the Institute (5.1–.2, 5.4).

With regard to the publication of details of the use of materials and results, the Institute discloses the genomic sequence data, including a description, into a freely accessible public domain accepted by the global scientific community (6.2, Schedule section 7), which is the Community Cyberinfrastructure for Advanced Microbial Ecology Research and Analysis (CAMERA) database. In addition, the Institute may publish articles which relate to the research in appropriate journals (6.3). For example, a complete journal issue about the second cruise can be retrieved online (JCVI, 2007). In any publications, the Institute is obliged to cite Australia as the source country and acknowledge the authorship of any Australian scientists who have provided significant advice or recommendations (6.4). In this regard, registration on the CAMERA website requires a user agreement, obliging the user to acknowledge the country of origin in any publication for which genetic information derived from the Expedition was used. The Institute is further obliged to provide regular progress reports, results, assessments of data, samples, assistance in interpreting the data and a final report within 90 days of the conclusion of the research (6.6–.9).

The Agreement contains additional provisions on confidentiality (7), liability (8), the commercial exploitation of material and published data (9), the termination of the Agreement (10) and dispute resolution (11).

Partnership for natural product discovery between the Griffith University, Queensland and AstraZeneca

From 1993 to 2007, the Griffith University's Eskitis Institute for Cell and Molecular Therapies formed a partnership with the pharmaceutical company AstraZeneca to identify natural bioactive molecules which could be potential leads for the discovery and development of new pharmaceuticals (Voumard, 2000, 7.21–.25; Jones, 2004, pp.115–44; SCBD, 2008, pp.41–53; Laird *et al.*, 2008; Australian Government, 2009, p.42). The original partnership was formed between Griffith University, the Queensland Pharmaceutical Research Institute and Astra Pharmaceuticals. Astra Pharmaceuticals merged with Zeneca in 1999 to become AstraZeneca. The Queensland Pharmaceutical Research Institute was renamed AstraZeneca R&D Brisbane, then evolved into the Natural Product Discovery Unit and finally moved under the aegis of the Eskitis Institute. The partnership was renewed in 1998 and 2002. Under the partnership, AstraZeneca invested over AUS$100 million of funding in the Eskitis Institute to participate in discovery and commercialization efforts. The involvement of the Eskitis Institute necessitated collection agreements with individual collection agencies. The bulk of collections took place within Queensland. Terrestrial samples were collected from rainforests by the Queensland Herbarium and marine samples were taken from the Great Barrier Reef by the Queensland Museum. Additional collections involved marine samples from Tasmania (conducted by Aquenal Pty Ltd.) and terrestrial samples from China (ZiYuan Medical Company), Papua New Guinea (Biodiversity Limited) and India (Biocon Ltd.). The collection of traditional knowledge was not included. The collection agencies then submitted their samples to the Eskitis Institute, which labelled each sample with its geographical location, made extracts of samples and analysed extracts against therapeutical targets of interest to AstraZeneca. Where an extract showed activity against a target, active compounds were identified and isolated for further analysis. For the duration of the partnership, AstraZeneca had exclusive rights over the samples and results from the analysis.

The collaboration resulted in the collection of over 45,000 terrestrial and marine samples. Marine collections involved around 9,500 samples from approximately 5,000 species of algae and invertebrates (sponges, cnidarians, bryozoans and ascidians), which are lodged at the Queensland Herbarium and Museum. The analysis of all extracts by the Eskitis Institute resulted in the discovery of over 800 bioactive compounds (200 from marine sources), which are currently under examination for potential commercial products.

In the course of the collaboration, many benefits accrued to the partners. The Eskitis Institute received annually AUS$7 million to conduct its research. In addition, royalties will be due should a successful product be marketed. With regard to non-monetary benefits, the Eskitis Institute extended its natural product expertise through close collaboration with and training received from AstraZeneca. This expertise has enabled the Eskitis Institute to begin new projects on natural product discovery with different partners. For example, the Eskitis Institute partnered with the Jarlmadangah Burru Aboriginal Community in Western Australia to develop pain-killing drugs from Aboriginal medicine.

In addition, ownership of samples remains with the Eskitis Institute, which allowed the Institute to build NatureBank™, a collection of 200,000 optimized compound extracts derived from the 45,000 samples. After the conclusion of the exclusive partnership with AstraZeneca, this collection has been made available for screening against therapeutical targets by different companies, thereby attracting new partners. For example, the Eskitis Institute formed a new partnership with Pfizer in 2008, under which Pfizer received access to NatureBank™. In return, Pfizer committed itself to spend over AUS$50 million on research and development initiatives in Australia and clinical trials as well as AUS$3.5 million for a Pfizer Research Fellows programme. In addition, Griffith University receives payments on drug development milestones and royalties from the marketing of anti-infective medicines. Similarly, Innate Pharmaceuticals received access to NatureBank™ to screen the database for novel drugs that block bacterial activities. In return, Griffith University will receive payments based on milestones and sales of medicines (EvaluatePharma, 2012; NewsMedical, 2012).

The Eskitis Institute also benefited from access to the latest technological equipment for natural product discovery and the publication of over 140 scientific articles written during the course of the partnership. Benefits for the Eskitis Institute also benefited Griffith University at large, because the technology is available to other university scientists, and a raised research reputation attracts the interest of new research partners.

The collecting agencies benefited from up-front fees per sample which covered the cost of staff, equipment and vehicles and also from potential royalties where commercial products are marketed. In addition, staff were trained in the taxonomic identification of samples, curatory and collection skills. Of special interest to the Queensland Herbarium and Museum is the increase of biodiversity information about new species and their geographical extent. For example, 3,000 new sponge species were sampled, of which 2,100 were new to science, resulting in a threefold increase of Australian sponge diversity extrapolations to 5,000 (global: 15,000). Taxonomic identification also proved useful to biodiversity conservation through the identification of endangered species and biodiversity hotspots (Myers, 1988, p.187). These results provide guidance for environmental planning and management of, for example, marine protected areas in the region.

Queensland benefited from AstraZeneca's financial investment, scientific and technical employment, an increase in their scientific and technological capacity, an enhanced research reputation and better knowledge of domestic biota. Furthermore, the Queensland Biodiscovery Act 2004 can also be regarded as a benefit, because experiences encountered during the partnership were incorporated into the drafting process of the Act. For example, taxonomic duplicates of the partnership were lodged in the Queensland Herbarium and Museum from the beginning in 1993. The Act now prescribes in section 30(1) that collected animal material needs to be given to the Queensland Museum and plant material to the Queensland Herbarium.

The key benefit for AstraZeneca was access to the 45,000 samples during the partnership. Further benefits were working in a country with a robust legal system that provided legal certainty and clear ABS regulations, as well as financial incentives in the form of higher prices paid by Australia for drugs from AstraZeneca.

Remarks

The rough outline of these case studies provides valuable insights into the implementation of research projects under national ABS law.

ABS collaborations can be highly beneficial for all parties involved. In cases where collaboration involves commercial aspects, a combination of monetary and non-monetary benefits for the provider and provider state is best practice. Non-monetary benefits are a particularly invaluable investment for receiving future benefits. For example, an enhanced research reputation through the successful completion of a collaboration attracts new research partners. In addition, capacity building enables the provider state to conduct its own preliminary research on samples, thereby increasing their value and, ultimately, the amount of benefits returned to the provider state once samples are transferred.

In order to attract research partners in the first place, it is necessary to create a favourable research environment. A clear and transparent national ABS legislation providing legal certainty, provider flexibility and user friendliness, as is the case in Australia, is key to attraction and successful implementation of ABS collaborations (CBD ABS GTLE, 2008b, p.30).

Finally, the Australia–Venter collaboration illustrates well how provisions of UNCLOS can be integrated into an ABS agreement. As the activities by the J. Craig Venter Institute qualified as marine scientific research, the conditions imposed on the Institute complied with those prescribed within Article 249(1) of UNCLOS.

Conclusions

This chapter has analysed key principles and concepts relating to genetic resources, compared CBD and UNCLOS provisions regulating activities on marine genetic resources, examined the national implementation of these conventions and analysed recent research projects on marine genetic resources.

Marine genetic resources located in areas within national jurisdiction fall under the principle of sovereign rights over natural resources. This means that a state has the exclusive right to dispose freely of its marine genetic resources and to control the actions of foreigners using these resources. However, although sovereign rights confer strong rights as they are rooted in the concept of sovereignty, they are limited. These limitations are specified within provisions of UNCLOS and the CBD.

The most important key concept relating to genetic resources is utilization. As utilization triggers benefit sharing, clarifying the scope of this concept is imperative for the smooth implementation of the ABS regime. To date, only the Nagoya Protocol provides a legally binding definition, which broadly points towards research and development and biotechnology. There are non-binding catalogues of criteria and specific uses which offer additional guidance on the exact scope of utilization. Such catalogues cover activities that fall under three types: using the reproduction ability of biological systems; using biological molecules for specific purposes and managing biological information. It will be up to the parties negotiating ABS-agreements to decide which of the listed uses eventually triggers what type of benefit sharing.

At the international level, UNCLOS and the CBD exist in parallel, complementing each other and not conflicting. This can be seen by looking at the very similar rights and obligations that regulate activities on marine genetic resources as well as the precedence rules which give priority to UNCLOS provisions. Nevertheless, member states adhering to both conventions must be careful to implement them coherently. Some rights under one convention are very strong, while they are compromised by the other. If member states ignore the compromise, they could create a conflict between the two conventions at the national level. The precedence rules in the CBD particularly provide guidance for such situations.

The national implementation of both conventions within typical user states is still very fragmented. Comprehensive user measures are often missing, as are provider measures. Where legislation is identified, it is often incomplete and vague. Only very few countries have managed to adopt noteworthy ABS legislation, such as Australia has for access to genetic resources and Norway has for user measures. Nevertheless, several countries are making initial tentative approaches around user measures, frequently using their patent laws to oblige their users to disclose the source state.

Finally, there are successful case studies of commercial and non-commercial ABS-agreements, which illustrate that ABS is a viable concept. However, despite these positive case studies, the ABS regime does have several weak points which will be the focus of the next chapter.

Due to reasons of time and space, this chapter has not elaborated upon the complex relationship between international intellectual property law and the CBD. There are several documents which provide an insight into issues raised and possible solutions (WTO, 2001, para. 19; Dutfield, 2002; WTO, 2006a, 2006b; Lochen, 2007, pp.46–111; WTO, 2008; Greiber *et al.*, 2012, pp.39–41; WTO, 2012).

4   Weak points of the access and benefit sharing regime

This chapter explains how bilateral ABS transactions create a deficient ABS system.

Introduction

The preceding chapter illustrated that the current paradigm of a fully functioning ABS regime envisages mainly bilateral exchange in ABS transactions: a user approaches a provider state; negotiates an ABS contract in accordance with the domestic legislation of the provider state; collects, exports and uses the resource and returns a share of (non-)monetary benefits to the provider state.

This approach is problematic in three major ways. First, it is unjust because it neglects all other source states sharing same resource. Second, it is ineffective because provider states cannot control the whole value chain of a genetic resource. Third, it impairs research and development, at least as long as provider measures are the only instrument for controlling user compliance. While these issues remain unresolved, bilateral ABS transactions that rely on the measures of the provider state for implementation maintain an ABS regime that is unjust, ineffective and impairs research and development (Vogel, 2000, p.48; Winter, 2009, pp.19–27; Kamau, 2011a, pp.171–181).

Injustice

Due to the physical properties of seawater as well as the biological behaviour of many marine organisms, it is highly likely that many marine genetic resources, especially microbial species, have broad ranges and therefore straddle national boundaries. This is especially true for countries in the southern hemisphere, because they tend to be clustered together (CBD, 2006, para11(d)). As a result, the source state of marine genetic resources, i.e., the states supplying genetic resources collected from in-situ sources according to Article 2.4 of the CBD, is rarely a single state but more often a group of countries that shares the natural distribution of a specific species (Ruiz, 2007, p.3; Soplín and Muller, 2009, p.7).

However, bilateral ABS transactions between a provider state and a user ignore this plurality of source states: one provider state decides unilaterally whether or not to grant access and receives the full share of the benefits derived from utilization. This approach not only deprives all other source states from taking part in access negotiations, but they also miss out on any benefits derived from utilization (Schrijver, 1993, p.23; Rainne, 2006, p.322). Situations where one state monopolizes benefits that rightfully 'belong' to all states sharing a particular resource are highly unjust (Schroeder, 2000, p.55; Chishakwe and Young, 2003, p.9 ; Soplín and Muller, 2009, footnote 38; Winter, 2009, p.19; Kamau, 2011a, p.180) and may create transboundary conflicts involving competition between source states and an inequitable use of resources.

Competition and inequitable use

In cases where several countries share a common marine genetic resource, the possibility of receiving benefits derived from their utilization could create a competitive environment between those countries (Görg and Brand, 2000, p.391; Wolfrum *et al.*, 2001, p.37). This competition may take various forms, often with detrimental consequences for both source states and the environment.

First, in order to prevent potential users from turning to another state with the same genetic resource, or to attract the maximum amount of users, source states are compelled to adopt weaker conditions during access negotiations, such as a reduced amount of benefits to be returned (CBD WG-ABS, 2004, para. 119). A race to the bottom in access agreements between source states sharing similar ecosystems might emerge, driving down the price of accessing genetic resources to a nominal fee (Vogel, 1997). For example, under a bioprospecting agreement between a subsidiary of Monsanto, G.D. Searle and Co., and Washington University concerning the collection of Peruvian medicinal plant extracts, royalty payments were as low as 0.2 per cent (RAFI, 1994). Such situations do not only lower the bargaining power of the source party and thus undermine the 'mutually agreed terms' requirement by Article 15.4 of the CBD, but may also effectively invalidate any agreed benefit sharing.

Second, the necessity of lowering benefit sharing conditions in order to attract users may reduce the efficacy of ABS as a tool to ensure conservation and sustainable use (CBD WG-ABS, 2005a, p.23). Through lowering benefit sharing conditions, genetic resources lose potential value. Thus, there are fewer financial incentives to conserve habitats and use resources sustainably in order to maintain the potential for the discovery of unknown genetic resources. Granting more lucrative but also unsustainable, extractive uses of the environment, such as bottom trawling, seems more attractive, resulting in a deterioration of the environment and the compromising of the objectives of the CBD.

Third, competition may also affect the quality of national ABS legislation (Brand and Görg, 2001, p.22). In order to attract a maximum amount of users, states may choose to lower access standards and benefit sharing obligations. Such a situation again undermines the objectives of the CBD, because more users collecting genetic resources also increases pressure on the environment (Öhman, 2002, p.61; Winter, 2009, p.26).

Fourth, competition can also exacerbate difficult relationships between states and create conflicts (UNGA, 1982, Annex preamble 8; Halle, 2009, p.5). The Belize-Guatemala territorial dispute constitutes an extreme example (Perez *et al.*, 2009). Guatemala has claimed certain parts of Belizean land and sea territory, which has led to several problems. These mainly involve the harvesting of various natural resources from land and sea areas claimed by both countries, which has resulted in several armed conflicts between both countries in the past. Bilateral negotiations have failed to reach a satisfactory conclusion of the conflicts, and both countries are currently preparing to take their dispute to the International Court of Justice to resolve the conflict.

Sharing the benefits from the exploitation of a shared marine genetic resource with only the provider state creates an inequitable situation. It is a general obligation of all states to conserve biological diversity. One of the incentives to support this is the potential of financial gains from future discoveries of useful genetic resources (Flitner, 1998, p.147). If two states make similar efforts to conserve their shared ecosystem, but the benefits accrue unilaterally to the state where access occurred, then this situation is inequitable, because the other state is left empty-handed, despite its efforts to conserve biological diversity.

That the unilateral actions of provider states cause injustice, competition and inequitable use does suggest that multilateral cooperation of states sharing the same genetic resource could improve this situation.

Contemporary approaches towards increasing justice in ABS

The idea that managing shared resources requires cooperation between states sharing this resource is not new. Various UN forums have recognized that conservation and the harmonious exploitation of shared natural resources require a system of cooperation based on good neighbourliness, information exchange, prior consultation and equitable use (UNGA, 1970b, Annex principle 4; 1973, preamble 6 and paras 1–2; 1978, preamble 5–6; 1979, preamble 8 and paras 2–3), which has been reiterated by, for example, Principle 24 of the 1972 Declaration of the United Nations Conference on the Human Environment, Article 3 of the 1974 Charter of Economic Rights and Duties of States, the 1978 United Nations Environment Programme Environmental Law Guidelines and Principles on Shared Natural Resources and Principles 5, 7, 9, 12, 26 and 27 of the 1992 Rio Declaration on Environment and Development (Handl, 1978, pp.40–45; Adede, 1979, p.511; Schrijver, 1993, p.28; Boyle, 1999, pp.906–909; Paradell-Trius, 2000, p.95; Nanda and Pring, 2003, p.19; Birnie *et al.*, 2009, p.192).

The International Court of Justice reached similar conclusions in its Fisheries Jurisdiction Case (*United Kingdom of Great Britain and Northern Ireland v Iceland (1974)*) stating that states wishing to exploit a shared resource must conduct 'negotiations in good faith for the equitable solution [of the dispute] …' taking into account 'the interests of other States in the conservation and equitable exploitation of these resources'. Today, many legal and policy regimes on the management of shared natural resources are either already in place for cases of, e.g., shared fish stocks (Article 63 UNCLOS and the 1995 United Nations Fish Stocks Agreement), transboundary freshwater resources (1997 Convention on the Law of the Non-Navigational Uses of International Watercourses) (Sadoff *et al.*, 2008) and large marine ecosystems (Sherman, 2005), or are in preparation, as in the cases of transboundary oil and gas fields (ILC, 2012).

Meanwhile, several instruments have either been put in place or are in development to manage genetic resources that cannot be attributed to a single source state: the Multilateral System of the Food and Agriculture Organization manages plant genetic resources; the CBD indicates a departure from pure bilateralism in ABS transactions; regional communities of states in the southern hemisphere are developing common legislation; and some legal authors have elaborated theoretical frameworks of regional organizations that could administer any genetic resources shared by two or more states.

The Multilateral System of the International Treaty for Plant Genetic Resources

The International Treaty on Plant Genetic Resources for Food and Agriculture aims to contribute to global food security through the conservation and sustainable use of plant genetic resources and sharing the benefits arising from their use (Article 1.1). To achieve this, the Treaty established the Multilateral System of Access and Benefit Sharing (10–13). The Multilateral System is a 'virtual and distributed genebank' of a predefined list of 64 genera of plant genetic resources (Governing Body, 2011a, para. 88–89). It maintains a central information repository of both the providers and users of these resources, establishes standard requirements for obtaining prior informed consent and entering into mutually agreed terms and manages a global trust fund. Any commercial benefits which users generate from using these resources flow into the fund, which then redistributes them worldwide to projects that contribute to global food security (Cooper, 2002, p.9; Tsioumani, 2004, p.28; Moore and Tymowski, 2005; Governing Body, 2006, 2011; Kamau, 2011b).

The Multilateral System justifies the global redistribution of funds by the special nature of plant genetic resources. They are special insofar as they can rarely be traced back to their original source, since a) agriculture in all countries depends on plant genetic resources that have originated elsewhere and b) the exchange of plant genetic resources has existed for millennia (Moore and Tymowski, 2005, p.79). Thus, they constitute global resources.

However, most genetic resources, such as marine genetic resources and those used for purposes other than food and agriculture, do not share this special global nature. First, while research and development concerning plant genetic resources is essential for supporting national economies and global food security, the utilization of marine genetic resources plays a minor role in these contexts. Second, whereas plant genetic resources have been exchanged globally for millennia, the research and development of marine genetic resources only began as recently in the 1950s. Furthermore, their exchange is mostly one-directional and most marine genetic resources are still undiscovered. As a result, most marine genetic resources theoretically have a clear origin. Thus, a global system, such as that implemented under the Multilateral System in its current form, would be inadequate for managing ABS of marine genetic resources.

The Convention on Biological Diversity and Nagoya Protocol

Although the CBD does not anticipate the possibility that genetic resources may straddle boundaries, some provisions do indicate a departure from straightforward bilateralism in ABS transactions. The CBD stresses the international and regional cooperation of states with regard to conservation and sustainable use in general (preamble 15); obliges parties to cooperate on matters of mutual interest (Article 5) and recognizes the sovereign rights states have over their genetic resources (15.1). By referring to 'their' genetic resources, the CBD does not imply ownership but geography (Glowka *et al.*, 1994, p.76). This provision can be interpreted to convey shared sovereign rights as soon as the geographical distribution of a genetic resource straddles national boundaries. In other words, states sharing the same genetic resources have a common responsibility to manage those resources, and this cannot be fulfilled by unilateral actions alone.

The Nagoya Protocol provides more substantial content in this regard. The Protocol prescribes that in 'instances where the same genetic resources are found *in situ* within the territory of more than one Party, those Parties shall endeavour to cooperate … with a view to implementing this Protocol' (8.1). Although this provision constitutes a clear derogation of absolute state sovereignty as advocated above, it suffers from three drawbacks. First, it expresses only a weak obligation ('endeavour') and is very broad. Second, transboundary cooperation can take a multitude of forms, many of which would not contribute to more justice. Third, it refers to the 'territory', which raises questions about geographical scope. The term 'territory' commonly applies to the land territory, internal waters and territorial sea but excludes the EEZ and the continental shelf (Kaczorowska, 2010, p.265). This interpretation contradicts the geographical scope of the CBD, which includes the EEZ and the shelf. Similarly, negotiation protocols of the Nagoya Protocol indicate that these zones should be included (CBD WG-ABS, 2007, p.41). If the geographical scope of this provision covers the EEZ and the shelf as well, it causes friction with UNCLOS. States may interpret 'territory' to apply to these zones as well and therefore claim full sovereignty in the EEZ and on the shelf. This conduct would undermine substantive provisions under UNCLOS and therefore contradict the CBD, which calls for the implementation of the CBD consistent with the rights and obligations under UNCLOS.

Regional communities

Many states of the southern hemisphere have recognized the need for multilateral cooperation and have developed regional legislation on the management of shared genetic resources. These include Decision 391 of the Andean Community (ten Kate, 1997; Correa, 2004, p.154; Bucher, 2008, p.110), the draft ASEAN

Framework Agreement on Access to Biological and Genetic Resources and the draft Framework Agreement of the Hindu Kush-Himalayan Region (Oli *et al.*, 2010). These commonly prescribe that: a) states sharing the same genetic resource must notify each other and take into account each other's interests when negotiating access; b) benefits should be shared among member states through a common fund and c) monitoring should mainly rely on the compliance of the user. The major drawback of these instruments is that they do not explain how to identify source states sharing the same genetic resources.

Gene cooperatives and the Biodiversity Cartel theory

Some experts have proposed the establishment of gene cooperatives or biodiversity cartels (Reid *et al.*, 1995, p.8; Vogel, 2000, 2007a, p.50, 2007b, p.124). These would aim to create multinational organizations that could negotiate access to genetic resources and traditional knowledge. As a corollary, the bargaining power of provider states would increase, which would ensure that prices for using genetic resources would stay above a reasonable threshold, something which would not occur in a competitive environment with multiple individual providers. These organizations would maintain central repositories to store and distribute genetic resources and traditional knowledge to users. If the utilization of genetic resources or traditional knowledge generated benefits, these benefits would accrue to all parties holding same resource or knowledge. Despite these advantages, the cartel theory in particular has been criticized because it constitutes an unviable approach for the effective management of genetic resources and traditional knowledge, since a) the fee-based services provided by the cartel often do not relate to genetic resources or traditional knowledge; b) the royalty rate is too high; c) the implementation of technology transfer is difficult; d) the focus lies on monetary benefits only; e) the scope of transaction costs is unclear and f) the competition with pre-CBD ex-situ collections is unresolved (Wolfrum *et al.*, 2001, p.38).

Conclusions

All of the above examples show that a more just ABS system would simply require the transboundary cooperation of source states when negotiating access or sharing benefits. However, prescribing transboundary cooperation alone will not result in a more just ABS regime, as long as the geographical distribution of a particular species remains unclear. While the distribution is unknown, several uncertainties would endanger the functioning of any ABS regime: which states should collaborate during access negotiations, and which states should receive benefits? To answer such questions requires instruments that allow the clear identification of source states.

In addition, the transboundary collaboration of source states alone does not assure effective monitoring once a genetic resource or knowledge has left the provider state.

Ineffectiveness

To date, 48 states have adopted ABS-relevant legislation (CBD, 2012). Most legislation focuses on provider measures. However, provider measures alone cannot ensure effective monitoring and compliance along the whole chain of valorization (both commercial and non-commercial). States can still have problems enforcing their rights in cases of non-compliance once a resource has left the country.

Monitoring is mainly hampered by the complexity of the valorization chain. It is a fallacy that the utilization of genetic resources always involves two countries: the source country and the user country (Tvedt and Young, 2007, p.2). The value-creation chain from an accessed genetic resource through to a final product involves a number of diverse steps and parties around the world, and numerous transactions (CBD ABS GTLE, 2008b, p.9). Many public entities, such as laboratories, universities, ex-situ collections and individual researchers, pursue a policy of open access to collections – over half a million microbial strains are exchanged informally each year (Dedeurwaerdere *et al.*, 2009, pp.7 and 37). In order to advance scientific progress and cooperation, they exchange material on an informal basis without contractual obligations (Biber-Klemm *et al.*, 2010, pp.5 and 16). These public entities often also serve as intermediaries. They collect and analyse material, thereby creating knowledge and information which can be useful for subsequent research and development (CBD EP-ABS, 1999, para. 16). To access this, many large companies form partnerships with public research organizations in or outside the source country (Reid *et al.*, 1993, p.25). These companies obtain interesting leads from natural products research conducted by their trade partners, rather than collecting it for themselves in the original provider state (CBD WG-ABS, 2005b, p.10; Holm-Müller *et al.*, 2005, p.43). As a result, the same material can be used by different parties for different purposes in different sectors (Mugabe *et al.*, 1997, p.12; Afreen and Abraham, 2008, p.25). A genetic resource gets modified, changes hands frequently and may result in products that have little similarity to the original resource once accessed (CBD ABS GTLE, 2008b, p.10; Biber-Klemm *et al.*, 2010, p.16). The chain of valorization turns into a complex web of valorization.

Additional reasons that impair effective monitoring are that research for commercial purposes is often kept secret until successful patenting (CBD ABS GTLE, 2008b, p.10), and reporting duties by users are enforceable only with difficulty (Kamau, 2011a, p.178).

Assuming that cases of non-compliance can be discovered, provider states face additional difficulties in enforcing their rights. First, provider state law cannot be enforced after genetic resources have been taken outside the country (Tvedt and Young, 2007, pp.16 and 49). The reach of a provider state's law and legal processes is coterminous with its territory (territory principle) (Winter, 2009, p.19). The main legal instrument for enforcing rights is a contract between provider and user. Second, if a contract is in place and it empowers provider state courts to deliver judgments, the execution of the judgment still relies on the courts of the user state (Kamau, 2011a, p.178). Third, providers may encounter difficulties in obtaining access to foreign justice (Normand, 2004, p.138). Fourth, the burden of proving non-compliance lies on the provider state. Fifth, the provider state carries the majority of monitoring and enforcement costs (Winter, 2009, p.19).

In addition, limited awareness, legal uncertainties, a lack of experience and low administrative capacities in most provider states do not only result in imprecise ABS legislation but also make monitoring and enforcement an insurmountable obstacle in tackling non-compliance (Pisupati, 2008, p.1; UNU-IAS, 2008, p.82; Stoll, 2009, p.13).

In conclusion, many provider states are incapable of controlling the utilization of their genetic resources once they have been exported (Biber-Klemm, 2008, p.17). They can neither monitor compliance nor enforce their rights in cases of non-compliance. The idea that provider measures alone can guarantee a globally effective ABS regime is misleading (Medaglia and Silva, 2007, pp.11 and 16).

Hampering research and development

There is a complex network of factors which have a negative impact on the conduct of research and development. First, effectively controlling the utilization of genetic resources after export by provider measures alone is virtually impossible. However, many states have been reluctant to adopt user measures. In fact, hardly any state has introduced legislative, administrative or policy measures obliging their users to comply with access and benefit-sharing conditions (Young, 2004, pp.5 and 76; 2008, p.117; Kamau *et al.*, 2010, p.256). Where states have adopted user measures, they are often far from comprehensive. Only Norway has adopted a number of user provisions. Most other states only have one provision at most, with many having none. If developed states have adopted ABS legislation, it primarily focuses on the provider side (Tvedt and Young, 2007, p.11). The main reason for this is that states fear that user measures will impose high costs and legal uncertainty on their users and thus curb their scientific and industrial performance (Rosendal, 1994, p.90; Tvedt and Young, 2007, p.16). Second, there have been cases where the ownership of genetic resources was not respected: resources were exported, developed and commercialized without obtaining the consent of the provider state or sharing the benefits (Biber-Klemm and Martinez, 2009, p.6; de Jonge and Louwaars, 2009, p.42). A recent case is Nestlé's rooibos 'robbery' in South Africa. Nestlé collected samples of rooibos and honeybush and filed various patents without the consent of providers and only entered into benefit-sharing negotiations after much public pressure (Berne Declaration, 2010). Such examples exacerbate provider states' fears that they may fall victim to similar biopiracy (Shiva, 1997).

Consequently, many states have begun to distrust users and the ABS principle (Klein and Düppen, 2008, p.45). They are concerned that their rights and interests will not be respected once genetic resources have been accessed and are being used in foreign countries (Normand, 2004, p.137). They fear that ABS might just be a modern tool to conceal the 'plunder by the wealthy industrialized western world' (Lindeskog, 2004, p.189).

The ineffective control given by provider measures, apathetic user states, the fear of biopiracy and the mistrust of users all combine to propel provider states to adopt overly restrictive regulations and obstacles to access (Medaglia and Silva, 2007, p.16; Tvedt and Young, 2007, p.17; Biber-Klemm and Martinez, 2009, p.17; Stoll, 2009, p.13). In other words, provider states see prohibitive access legislation and tight contract conditions as the only instruments by which they can prevent illegitimate use of their resources and to enforce benefit sharing. Users have a number of common criticisms of access legislation, including: a) cumbersome application procedures which involve long periods of time from filing the application to final approval of a contract; b) prior informed consent requirements which are administratively tedious and burdensome; c) very high transaction costs to obtain consent from all relevant stakeholders; d) complex institutional mechanisms, e.g., decision-making procedures involving all relevant stakeholders; e) unrealistic benefit-sharing expectations; f) prohibitions on using the resource for purposes other than that requested; g) restrictions on transfer to third parties; h) restrictions on obtaining intellectual property protection and i) that the same highly restrictive conditions apply to both academic researchers without commercial intentions and users with commercial intentions (CBD WG-ABS, 2005b, p.30; Medaglia and Silva, 2007, pp.8–12; Dedeurwaerdere *et al.*, 2009, p.37).

The consequence of such overly restrictive access legislation is a two-fold chilling effect on research and development (CBD WG-ABS, 2005b, p.38). First, users must invest much time and money as they attempt to penetrate the law and obtain permission to operate (Miller, 2007, p.66). Academic researchers, who generate most biodiversity knowledge, will have particular difficulties when access legislation does not consider the special circumstances of academic researchers: They often work under adverse conditions, are chronically underfunded and do not have the financial flexibility or legal capabilities to negotiate complex access contracts with governmental institutions (Grajal, 1999, p.7). As a result, collecting, inventorying and monitoring biodiversity are discouraged. Second, many researchers have begun to withdraw from the free exchange of material between peers, pursuing a more restrictive formal exchange. This further limits the amount of material that is available and usable by the global research community (Dedeurwaerdere *et al.*, 2009, pp.11 and 37).

The overall effect is that strict national ABS legislation deters the very type of research that the CBD should stimulate (CBD ABS GTLE, 2008b, p.7). If a law is too cumbersome, fewer genetic resources will be sought, less biodiversity knowledge will be generated, fewer benefits will be created, and opportunities for conservation and sustainable use will be lost (Grajal, 1999, p.8; Jeffery, 2002, p.750; Dávalos *et al.*, 2003, p.1518; Pisupati, 2007, p.34; Biber-Klemm *et al.*, 2010, p.5). In short, strict access legislation undermines the objectives of the CBD.

Conclusions

This chapter has illustrated that bilateral ABS transactions which rely exclusively on measures taken by the provider state cannot ensure a just and effective ABS regime and in fact can even discourage research and development.

First, such transactions and measures rarely acknowledge that marine genetic resources frequently straddle national boundaries. In such cases, it would be unjust if the provider state disregards the interests of the other source states and claims the whole share of benefits derived from utilization. But even where cooperative measures are in place, they do not create a wholly just system, because they lack the instruments to show the geographical distribution of species. As long as the distribution of a resource remains unknown, the true number of source states will remain unidentified.

Second, provider measures alone cannot ensure that the utilization of genetic resources is monitored, nor can they ensure the compliance of users. Once a user exports a genetic resource, the provider state often loses track of its fate. A genetic resource can be exchanged and modified and may result in products that cannot be traced back to the source states. Thus, products resulting from resources cannot be attributed to their source states, which must unwittingly forgo any benefits derived from utilization. This makes the ABS regime highly ineffective, at least as long as there are no instruments that can illuminate and clarify the process of valorization.

Third, ineffective control by provider states, apathetic user states, the fear of biopiracy and the mistrust of users by providers all combine to drive provider states to adopt overly strict ABS obligations. This hampers research and development, because users: a) have problems accessing resources and b) refrain from freely exchanging material among themselves in order to avoid accusations of breaching ABS obligations.

These problems illustrate that provider measures alone are insufficient to create a just and effective ABS regime and to foster research and development (Medaglia and Silva, 2007, p.16; Tvedt and Young, 2007, p.16). Instead, they show a real need for user side measures (Young, 2004, p.57; Stoll, 2009, p.13).

5 Biological databases for improving the ABS system

This chapter introduces biological databases and shows how they contribute to the ABS system.

Introduction

A fully functioning ABS system cannot rely on measures taken by the provider state alone and needs additional measures to be taken on the user side. The member states of the CBD have anticipated the limitations of provider measures and have discussed various user-side measures that could improve monitoring and ensure compliance. These include mainly checkpoints in user states and international recognized certificates of compliance as adopted by Article 17 Nagoya Protocol (Kamau *et al.*, 2010, p.252). However, member states have so far neglected the important contribution that biological databases could make to improving the ABS system: biological databases are available tools that could contribute not only to a more just and effective ABS system which facilitates research and development, but could also provide the scientific tools to implement the 'global multilateral benefit-sharing mechanism' for managing benefits derived from utilization of transboundary genetic resources, as envisaged by Article 10 of the Nagoya Protocol to the CBD.

This chapter: a) provides an overview of the scientific discipline behind biological databases (bioinformatics); b) analyses selected databases by using specific criteria and c) illustrates the potential contribution to justice and effectiveness by applying selected databases to a sample set of products from marine genetic resources.

Bioinformatics

Biological research generates tremendous amounts of data, which is often not published in peer-reviewed journals but is instead deposited in databases (Apweiler *et al.*, 2003, p.343). Data covers a wide range of biological information, mainly from molecular biology, and includes the sequences and structures of nucleic acids, proteins and other biochemical compounds as well as their functions and interactions. Other biological research fields also contribute data, such as biogeographical

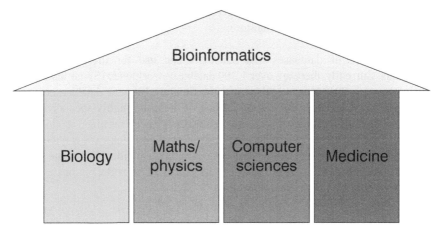

Figure 5.1 Disciplines contributing to bioinformatics

research or taxonomy. Bioinformatics broadly covers the application of hardware and software for gathering and storing these data in local or global, commercial or freely-accessible, online databases (Benoît and College, 2006, p.180).

Bioinformatics is by no means an insignificant science. It is an interdisciplinary field, which combines computer science, mathematics, physics, medicine and biology (Figure 5.1). Each of these disciplines may apply different standards, nomenclature, file formats and tools. Thus, integrating multiple datasets involves organizing very large and heterogeneous datasets and determining computer algorithms that cluster and display the data in biologically meaningful ways. In short, bioinformatics is a complex information management system for the vast field of biological research (Bayat, 2002, p.1018; Benoît and College, 2006, p.187).

Bioinformatics has many practical applications. Its main aim is to allow specialized scientists to access, visualize, analyse and interpret biological data. Through the application of bioinformatics, scientists can in particular (Goodman, 2002, p.68; Köhler, 2004, p.61; Benoît and College, 2006, pp.180–199):

- visualize gene and protein sequences, protein domains and protein structures (linear sequences, two- or three-dimensional models);
- compare nucleotide and protein sequences;
- reveal biological relationships between species (taxonomy and evolution);
- discover meaningful patterns or rules in nucleic or amino acid sequences;
- draw conclusions on gene expression patterns, protein interactions and metabolic pathways;
- predict the functions of genes, proteins and biochemicals and their interaction with disease pathways;
- understand the regulatory mechanisms of gene expression;
- obtain background information on chemical, physical and biological properties, on biotechnological application and on therapeutic and commercial uses;

- locate occurrences of species (biogeography); and
- access relevant scientific literature.

The number of databases that scientists can use for these purposes is enormous. Currently, there are over 1,300 databases worldwide (Swiss Institute of Bioinformatics, 2012a), and each year approximately 100 new databases emerge (Galperin and Cochrane, 2011, p.D1). Most of these are highly specialized and focus on specific organisms, disease pathways, protein families or domains, biomolecular tools or services. Despite this potential complexity, it is enough for this work when databases 'simply': a) connect biological molecules to the source species; b) connect the source species to its occurrences; and c) indicate which users have utilized a biological molecule.

Tracing products to their source countries and monitoring users

This section analyses the biological databases that could contribute to a more just and effective ABS system. To do this, databases must realize two goals. First, they must allow the tracing of products from marine genetic resources back to their source countries. Because no single database fulfils this task completely, this section divides databases into those that: a) link products from genetic resources to their source species and b) link the source species to its geographical distribution (Table 5.1). Second, they must enable the identification of both users and the uses of genetic resources.

From the vast number of biological databases available, this section analyses only few databases that fulfil specific criteria. Those databases on biological molecules must: a) include molecules from *marine* sources; b) refer to the source species and c) be global in their approach. Likewise, databases on geographic distribution must: a) unequivocally identify the source state by mentioning coordinates or the source states; b) include marine species and c) be global. The following sections and Table 5.2 at the end of the following analyses examine the databases in more detail by using selected criteria. These criteria involve:

Table 5.1 Biological databases analysed in this chapter

Product – species	*Species – geography*
Dictionary of Marine Natural Products	Ocean Biogeographic Information System
Thieme RÖMPP Online	Global Biodiversity Information Facility
The Merck Index	
Natural Products Alert	
PubChem	
GenBank	
UniProt	
The NCBI Protein Database	

- type and form of stored information;
- usability of the database;
- legal form (public, private);
- rules regarding feeding and retrieving data;
- monitoring opportunities;
- data limitations; and
- ABS elements (intellectual property, model contracts, benefit sharing, etc.).

Linking products from marine genetic resources to the source species

This section introduces commercial and non-commercial databases that link biological molecules to their source species. The databases cover a broad range of molecules including biochemical compounds, nucleic acids and proteins.

Dictionary of Marine Natural Products

The Dictionary of Natural Products is a comprehensive database holding information on almost every known chemical substance from natural sources and is available over the internet, as a book and on CD-ROM. It has been compiled over the past 25 years by a team of academics and freelancers who have gathered descriptive and numerical data on the properties of chemical compounds by reviewing the natural product literature (CHEMnetBASE, 2012a). The Dictionary of Marine Natural Products, which will be analysed here, is a subset database containing additional information on chemical substances from marine sources only (Blunt and Munro, 2008).

TYPE AND FORM OF STORED INFORMATION

The database contains more than 35,000 entries on compounds belonging to different structural classes (carbohydrates, terpenoids, steroids, aminoacids, peptides, alkaloids, etc.). Each entry comprises a parent compound and closely related compounds such as derivatives and variants, where available. Each compound is described by 35 fields covering: systematic and common names; structural formulas; literature references; as well as chemical, physical and biological properties. Biological properties include the biological use or importance, such as a particular medical application and the biological source, identified by the source species, where possible (Box 5.1).

USING THE DATABASE

Retrieving information on compounds requires the installation of a user interface program (CD-ROM version) or a browser plug-in (internet version). The interface provides a search form window, which is split into three panes (Figure 5.2). The 'structure search pane' allows the drawing of molecule structures for which the database can then be searched. The 'search terms pane' enables text queries from one or more of the 35 fields that describe the compounds. The 'index pane' displays all indexed terms within one field and lists matching terms while typing into the index pane. All queries can be conducted by using the Boolean operators AND, OR and NOT. The results are provided as a hit list of entries, giving the names of the compounds and molecular formula. Once an entry is opened, it can be viewed as indicated in Box 5.1 and exported as an Excel file.

LEGAL FORM

The Dictionary of Marine Natural Products ultimately falls under the management of the public limited company Informa, which is a multinational conference and publishing company (Informa, 2012). Informa publishes newsletters, academic journals, academic and business books, as well as commercial databases and is split into a professional and scientific division. With regard to the scientific division, Informa owns Taylor & Francis, which is an international academic publisher covering humanities, social sciences and natural sciences (Taylor & Francis Group, 2012). Taylor & Francis is, in turn, divided into two main branches covering journals and books. On the book side, Taylor & Francis owns CRC Press, which is a publishing group specializing in technical books related chiefly to engineering, mathematics and science (CRC Press, 2012). As well as publishing printed material, CRC Press has built an online presence through CRCnetBASE, providing e-books as well as subscription-based databases (CRCnetBASE, 2012). A collection of these databases is available under CHEMnetBASE, which is a conglomeration of various databases, including the Dictionary of Marine Natural Products (CHEMnetBASE, 2012b). All databases are available for a fee. A university site licence for the Dictionary of Marine Natural Products amounts to US$595 annually.

RULES CONCERNING FEEDING AND RETRIEVING DATA

The staff of the Dictionary of Marine Natural Products review relevant publications, and, as such, there is no active feeding of data. However, researchers can highlight their publication to the staff.

To retrieve data, users must abide by a number of rules. First, in order to use the online version of the Dictionary, users must register and provide personal contact details and financial information. Second, because CRC Press is the owner of the information within the database, users must cite the Dictionary when extracting and using information for their purposes. Third, users must pay a certain amount of money before they can access the information within the database.

MONITORING OPPORTUNITIES

In principle, the Dictionary allows the users of genetic resources to be monitored via its reference list. It contains a selection of central scientific papers at the end of each compound entry, so that users can find additional information on any particular compound. However, the value of using these reference lists to monitor users is questionable. First, the Dictionary does not give any indication about current uses of the resource, because it only refers to publications, i.e., the presentation of results from concluded activities. Second, each entry cites only a selection of central papers, which does not represent the full spectrum of users and uses. Third, by referencing only scientific papers, the identification of commercial uses is precluded. Finally, the Dictionary does not provide the papers themselves, which can often be only accessed for a fee.

DATA LIMITATIONS

Adding to the limitations concerning monitoring, the Dictionary also suffers from other drawbacks related to the completeness of the database. The authors of the Dictionary have acknowledged that it is an unrealistic target to list all compounds that occur in the sea (Blunt and Munro, 2008, p.xv). They highlight particular classes of compounds that are underrepresented within the database. These include:

- biochemicals endogenous to higher marine animals;
- microbial products isolated from organisms with widespread dispersal on both sea and land;
- natural products of a 'terrestrial' type isolated from plants and animals in coastal environments;
- widespread polysaccharides; and
- lipids.

Most of these limitations are covered by the parent database Dictionary of Natural Products, which was not available for this work.

ACCESS AND BENEFIT-SHARING ELEMENTS

The Dictionary itself is indirectly involved in the utilization of genetic resources, because it publishes data from direct utilization obtained from the scientific literature for a fee. Having said this, there are no references to ABS in the Dictionary. CRC Press is the sole owner of the information (Dictionary of Marine Natural Products, 2012).

Thieme RÖMPP Online

RÖMPP Online is a large, fee-based chemical encyclopaedia in German maintained by a team of editorial staff and over 250 authors who are experts in their various scientific fields (Füllbeck *et al.*, 2006, p.351; Thieme Chemistry, 2012a). It contains over 60,000 entries on chemistry, biotechnology and genetic engineering, food chemistry, environmental and process technology, as well as natural products. With regard to the latter, the encyclopaedia contains over 5,000 entries on approximately 6,000 important natural products (Steglich *et al.*, 2000). This section only analyses online data on products provided by marine organisms.

TYPE AND FORM OF STORED INFORMATION

Although 5,000 entries on natural products may seem an impressive number, RÖMPP online holds only 313 entries on marine natural products. This can be explained by its explicit focus on only the most important natural products.

Box 5.1 Selected elements of an entry from the Dictionary of Marine Natural Products, using spongothymidine as an example. The biological source is highlighted. Source: Blunt and Munro (2008)

Entry name:	1-Arabinofuranosylthymine
Synonym(s):	Spongothymidine. Ara-T
Chapman Hall number:	NKG75-G
CAS Registry number:	605-23-2
Type of compound code(s):	ZS4000 VE9900 XA2400
Molecular formula:	$C_{10}H_{14}N_2O_6$
Molecular weight:	258.23
Accurate mass:	258.085188
Percentage composition:	C 46.51%; H 5.46%; N 10.85%; O 37.17%
Biological source:	Isol. from Caribbean sponge, *Cryptotethia crypta*
Biological use/importance:	Antiviral agent. HIV reverse transcriptase (HIV-rt) inhibitor
Melting point:	Mp 194–195°, Mp 242°
Optical rotation:	$[\alpha]^{28}_D$ +90 (c, 0.8 in Py)
UV:	[base] λ_{max} 267 () (pH 12) (Derep) [neutral] λ_{max} 267 (ε 9600) (pH 7.2) (Derep) [neutral] max 269 (ε 9250) (H2O) (Berdy)
Other data:	Low acute toxicity
RTECS accession number:	XP2100200

References:

Bergmann, W. *et al.*, J.O.C., 1955, 20, 1501–1507 (isol, struct)

Fr. Pat., 1965, 1 396 003, (Upjohn); CA, 63, 13392d (synth)

Tougard, P., Acta Cryst. B, 1973, 29, 2227–2232 (cryst struct)

Mueller, W.E.G., FEBS-Symp., 1979, 57, 327–341 (rev)

Ooka, T. *et al.*, Virology, 1980, 104, 219–223 (activity)

Soike, K.F. *et al.*, Antiviral Res., 1984, 4, 245–257 (pharmacol, activity)

Gosselin, G. *et al.*, Nucleosides Nucleotides, 1984, 3, 265–275 (synth)

Machida, H. *et al.*, Microbiol. Immunol., 1991, 35, 963–973 (pharmacol)

Like the Dictionary of Marine Natural Products, the encyclopaedia characterizes each natural compound by a comprehensive set of 64 attributes described in concise texts, tables and figures. These attributes include, where available: the structural formulas of parent compounds, derivatives and similar compounds; physicochemical data; isolation and synthesis; biological activity; biotechnological application; relevant literature; relevant legal provisions on utilization and the biological source in Latin species names (Box 5.2) (Deckwer *et al.*, 2005, p.23).

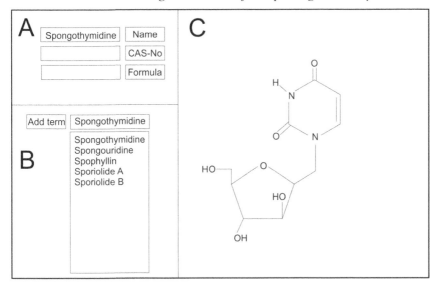

Figure 5.2 Sketch of the search window of the Dictionary of Marine Natural Products (CD-ROM version) using spongothymidine as an example. The search terms pane is located on the top left (A), the index pane on the bottom left (B) and the structure search pane on the right side (C).

Source: Blunt and Munro (2008)

USING THE DATABASE

Accessing RÖMPP Online requires only an internet browser, without the installation of any programmes or plug-ins. The search pane allows intuitive inquiries using either simple keyword searches or advanced searches for keywords in combination with other criteria, e.g. Chemical Abstract Service (CAS) registry numbers or topics, such as 'marine organisms'. The database provides a live hit list of exact and similar matches from which the user may choose (Figure 5.3).

LEGAL FORM

Thieme is a German limited publishing partnership for journals and books, focusing primarily on all branches of human medicine (Thieme, 2012). Other branches include biology and chemistry. Thieme Chemistry publishes comprehensive information on scientific fields related to chemistry within journals, reference works, monographs and encyclopaedias (Thieme Chemistry, 2012b). Many of these products are available in printed form, on CD-ROM and online, such as the RÖMPP encyclopaedia.

RULES CONCERNING FEEDING AND RETRIEVING DATA

To feed data into the encyclopaedia, the authors compile and verify relevant information on specific compounds and submit text-based files. They are experts in their scientific fields who track progress in the development of new natural

Box 5.2 Elements of an entry from RÖMPP Online using ecteinascidin as an example (RD-05-00096). Information on the biological source is highlighted. Source: Thieme Chemistry (2012a)

Aus der karib. Mangroven-Ascidie *Ecteinascidia turbinata* wurden antineoplast. wirksame Tris-tetrahydroisochinolin-Alkaloide isoliert, die E., z. B. *Et 743* ($C_{39}H_{43}N_3O_{11}S$, M_r 761,84), aktiv gegen L1210-Zellen (IC_{50} 0,5 ng/mL) und P388-murine Leukämie. Die E. sind eng verwandt mit den Renieramycinen aus dem Schwamm *Reniera* sp., z. B. *Renieramycin A*, $C_{30}H_{34}N_2O_9$, M_r 566,61, $[\alpha]_D$ −36,3 (Methanol). Et 743 ist in klin. Entwicklung (Phase II) gegen Lungen-, Knochen-, Dickdarm-, Brust-, Eierstock- und Nierentumore sowie gegen Melanome und Weichteilsarkome [1].

Übersetzungen: Ecteinascidines

CAS RN: 114899-77-3 (Et 743)

Literatur: [1] R & D Focus, Drug News, 22.2.1999, S. 7; J. Med. Chem. **42**, 2493 (1999); Proc. Natl. Acad. Sci. USA **96**, 3496 (1999); J. Am. Chem. Soc. **104**, 265 (1982), 9017 (Biosynth., abs. Konfiguration), 9202 (1996) (Synth. Et 743) ; **119**, 5475 (1997) NMR des DNA-Addukts); J. Org. Chem. **54**, 5822 (1989); **55**, 4508, 4512 (1990); Nachr. Chem. Tech. Lab. **41**, 6 (1993); Org. Lett. **1**, 75 (1999) (Wirkung); Proc. Natl. Acad. Sci. USA **89**, 11456 (1992); Synlett **1999**, 1103; Tetrahedron Lett. **33**, 3721 (1992).

---

products and integrate these into the database. At the same time, they review changes in existing natural products and update the database accordingly (Thieme, 2010, p6).

Data can be accessed after the conclusion of a licence agreement with Thieme. Under the agreement, the user has to pay an annual licence fee, which depends on the number of users and the type of the user institution, e.g., university, school, public authority or private company. Once the agreement has been concluded, users can access the data via their account.

MONITORING OPPORTUNITIES

Monitoring opportunities in RÖMPP Online are limited. Like the Dictionary of Marine Natural Products, RÖMPP Online only provides a reference list of the central scientific publications where additional information can be found. The paper itself cannot be accessed.

DATA LIMITATIONS

While the great strength of the database is the very broad coverage of various scientific fields, it is necessarily weaker in other areas. For example, the section on marine

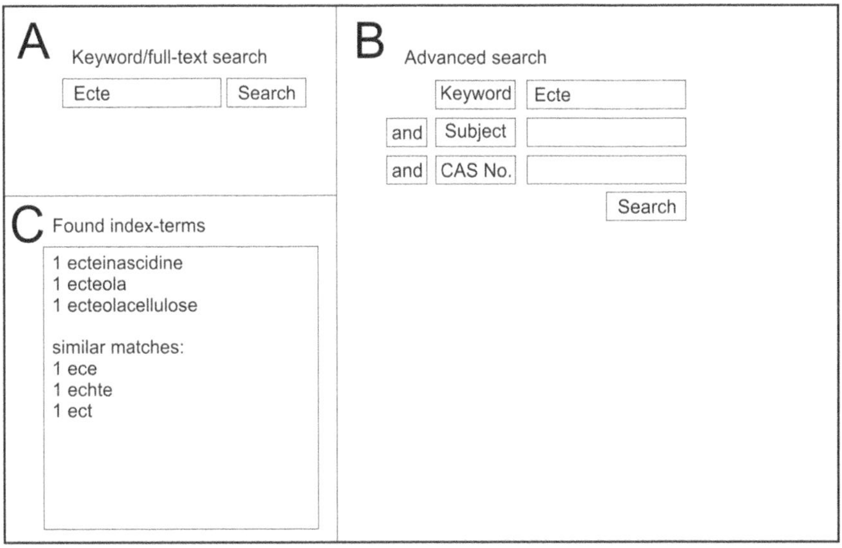

A Keyword/full-text search

| Ecte | Search |

B Advanced search

Keyword | Ecte

and | Subject |

and | CAS No. |

Search

C Found index-terms

1 ecteinascidine
1 ecteola
1 ecteolacellulose

similar matches:
1 ece
1 echte
1 ect

Figure 5.3 Sketch of the RÖMPP search pane using ecteinascidin as an example. Typing the first letters of 'ecteinascidin' in the simple keyword search field (A) or the advanced search field (B) yields a hit list of exact and similar matches (C) from which the user may choose.

Source: Thieme Chemistry (2012a)

natural products contains only about 300 compounds. RÖMPP Online explicitly aims to list only the most important compounds, e.g., those that have potential applications, but it omits most of their other derivatives or variant forms. In addition, entries vary greatly in length, depending on the amount of relevant information published for the compound, and the source species is not always mentioned.

ACCESS AND BENEFIT-SHARING ELEMENTS

Services provided by RÖMPP Online on natural products constitute an ABS constellation, because information on natural products extracted from source countries is reproduced in return for a user fee. However, no ABS elements were identified during analysis of the database or the sample licence agreement. The content of the encyclopaedia is protected by copyright.

The Merck Index

The Merck Index is a commercial one-volume encyclopaedia of important chemicals, drugs and other biologically active molecules, covered in more than 11,000 monographs (entries). Of these, approximately 6,000 cover drugs, pharmaceuticals, naturally-occurring substances and plants. It is available on the internet (CambridgeSoft, 2012) and as a book, including a companion CD-ROM (O'Neil *et al.*, 2006). This section analyses the latter.

TYPE AND FORM OF STORED INFORMATION

The monographs vary greatly in length, depending on available information. Each monograph can include information on: the compound and its derivatives; common, trade and systematic names; trademarks and associated companies; registry numbers of the CAS; physical and toxicity data; therapeutic and commercial uses; scientific and patent literature; chemical structures and formulas; molecular weights and the biological source, where available (Box 5.3).

USING THE DATABASE

As soon the database is installed, the user needs to activate the encyclopaedia using a unique serial number. A user then accesses monographs via a search screen interface, which allows text and property searches. Text searches include compound names, CAS registry number, keywords, manufacturer, monograph title and (non-) therapeutic uses. Property searches include molecular formulas, mass, various physical data and toxicity (Figure 5.4). Queries only return exact matches, which makes exact spelling crucial for successful results.

In addition, the search screen can be upgraded for a fee, enabling queries for identical or similar molecular structures.

LEGAL FORM

The Merck Index is published by the large pharmaceutical company Merck & Co, Incorporated. Apart from discovering, developing, manufacturing and marketing products for both human and animal health, Merck & Co also publishes a series of medical books.

RULES CONCERNING FEEDING AND RETRIEVING DATA

Merck staff review relevant primary and secondary literature sources and assemble data on monographs before feeding data into the Index.

The Merck Index CD version can be purchased for an academic (US$190) or a business price (US$490). Before installing the CD version software, the user has to enter into a licence agreement. Under the agreement, the user must acknowledge the Merck Index as the source of information. The user is not allowed to: distribute copies to others; make the database publicly available, e.g., over the internet; or sublicense, rent, lease or lend any part of the database.

MONITORING OPPORTUNITIES

The Merck Index provides additional means for monitoring, besides reference lists of scientific literature. These include references to central patents in the United States and the World Intellectual Property Organization, as well as references to commercial trademarks and the relevant manufacturer. As a result, the Merck Index not only gives a broad overview of non-commercial uses, but also discloses commercial uses and users.

Box 5.3 Elements of a monograph from the Merck Index using ecteinascidin as an example. The biological source is highlighted. Source: O'Neil *et al.* (2006)

Monograph number:	0003503
Title:	Ecteinascidins
Literature references:	Family of tetrahydroisoquinoline alkaloids with antitumor activity isolated from the Caribbean tunicate, *Ecteinascidin turbinata*. Ecteinascidin 743 is the most abundant; bends DNA toward the major groove by selectively alkylating guanine from the minor groove. Prepn: K. L. Rinehart, T. G. Holt, **WO 8707610**; *eidem,* **US 5089273** (1987, 1992 both to Univ. Illinois); and structures: K. L. Rinehart *et al., J. Org. Chem.* **55**, 4512 (1990); R. Sakai *et al., Proc. Natl. Acad. Sci. USA* **89**, 11456 (1992). Crystal structures: Y. Guan *et al., J. Biomol. Struct. Dyn.* **10**, 793 (1993). Biosynthetic studies: R. G. Kerr, N. F. Miranda, *J. Nat. Prod.* **58**, 1618 (1995).
Derivative type:	Trabectedin
CAS Registry number:	114899-77-3
CAS name:	(1'R,6R,6aR,7R,13S,14S,16R)-5-(Acetyloxy)-3',4',6,6a,7,13,14,16-octahydro-6',8,14-trihydroxy-7',9-dimethoxy-4,10,23-trimethylspiro[6,16-(epithiopropanoxymethano)-7,13-imino-12H-1,3-dioxolo[7,8]isoquino[3,2-b][3]benzazocine-20,1'(2'H)-isoquinolin]-19-one
Additional names:	ecteinascidin 743
Manufacturers' codes:	ET-743
Trademarks:	Yondelis (PharmaMar)
Molecular formula:	$C_{39}H_{43}N_3O_{11}S$
Molecular weight:	761.84
Percent composition	C 61.48%, H 5.69%, N 5.52%, O 23.10%, S 4.21%
Literature references:	Enantioselective total synthesis: E. J. Corey *et al., J. Am. Chem. Soc.* **118**, 9202 (1996). *In vitro* antitumor activity: E. Isbicka *et al., Ann. Oncol.* **9**, 981 (1998). HPLC determn in plasma: H. Rosing *et al., J. Chromatogr. B* **710**, 183 (1998). Clinical pharmacokinetics: C. van Kesteren *et al., Clin. Cancer Res.* **6**, 4725 (2000). Review of mechanism of action: G. J. Aune *et al., Anti-Cancer Drugs* **13**, 545–555 (2002); of clinical development: C. van Kesteren *et al.,* ibid. **14**, 487–502 (2003); of metabolism and hepatotoxicity: J. H. Beumer *et al., Pharmacol. Res.* **51**, 391–398 (2005).

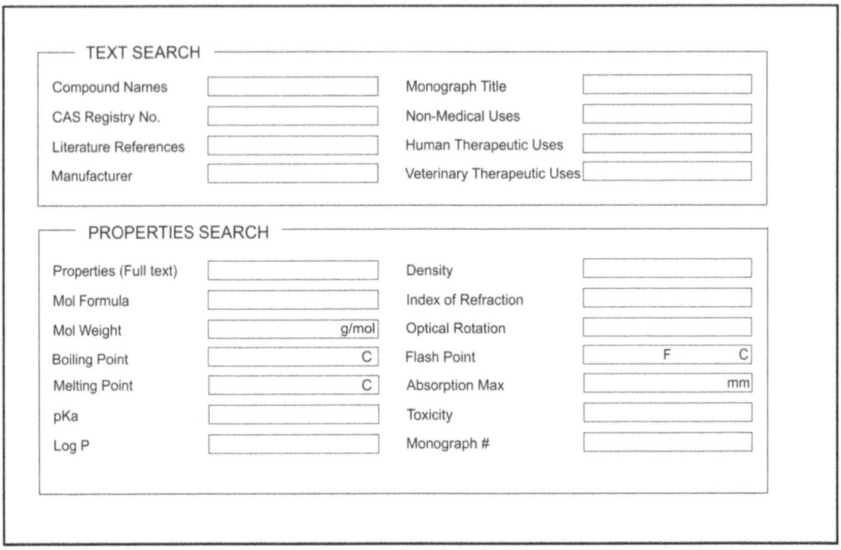

Figure 5.4 Sketch of the search screen interface of the Merck Index. Compound names can be monograph titles, CAS names, synonyms, trademarks, manufacturer codes and derivative codes. Molecular formula searches use the Hill Convention (C, then H then other elements in alphabetical order). Properties are range searchable. Temperatures in °C.

Source: O'Neil *et al.* (2006)

DATA LIMITATIONS

A comprehensive review of the limitations of the database is beyond scope of this work. The content of the database has been rated positively in one review, while the search interface, user support, cost (for the online version) and service were rated negatively (Californian Council of Chief Librarians, 2004). In addition, it is not clear how many compounds from marine source are covered, the biological source is not always stated and the literature list contains only central papers.

ACCESS AND BENEFIT SHARING ELEMENTS

As above, no explicit reference is made to any ABS elements. With regard to ownership, the database is the sole property of Merck & Co., Inc. The content of the database is protected by United States and international copyright laws, database rights and other intellectual property rights.

Natural Products Alert

The Natural Products Alert (NAPRALERT) database aims at including all natural products including: secondary metabolites; ethnomedical information; pharmacological and biochemical information; organism information and in vitro, in situ and in vivo and clinical trials. Currently, the database includes information from over 200,000 scientific papers and reviews (Loup *et al.*, 1985; NAPRALERT, 2012).

The database stores information in downloadable reports. Each report contains details of: source organisms; biological activities, such as anti-tumour, anti-viral or anti-inflammatory activity; ethnomedical applications, such as gastronomic, hypnotic or nervine purposes and relevant articles or reviews.

USING THE DATABASE

Detailed reports on compounds can be retrieved for a fee per citation. Reports that include 100 or fewer citations cost US$0.50 per citation, while reports with 101 or more citations cost US$0.25 per citation. In order to be able to generate reports, the user must create an account and provide contact information. A contract on terms and conditions does not have to be agreed. It is difficult for users not knowledgeable in the field to generate any successful hits, because the search algorithm not only requires exact spelling but also the full term; in other words, parts of terms are not recognized. Following verification, the user can search the database by organisms, compounds, pharmacological activity or authors. After a successful search has been completed, the user must submit credit card information and may then download the relevant report as a pdf.

LEGAL FORM

NAPRALERT is a registered service mark which ultimately falls under the responsibility of the University of Illinois at Chicago (University of Illinois at Chicago, 2012). The university is made up of a number of colleges, including the College of Pharmacy. This college in turn operates the Program for Collaborative Research in the Pharmaceutical Science, which is a research centre for the study of biologically active natural products. This maintains the NAPRALERT database and funds it primarily through individual access payments and voluntary donations.

RULES CONCERNING FEEDING AND RETRIEVING DATA

The database receives its information from systematic searches of all major journals pertinent to natural products performed by the database staff. Thus, there are no specific guidelines for data holders who wish to submit their data.

The retrieval of data is similarly unregulated. Apart from submitting private contact information and paying an access fee, users are free to use the database and data as they wish.

MONITORING OPPORTUNITIES

The monitoring potential of NAPRALERT is rather limited. It cites central research papers in the literature section of downloadable reports. Occasionally, the database may also refer to a patent.

DATA LIMITATIONS

There are deficiencies in the scope of the data. The literature from 1975 to 2003 is comprehensively analysed, while only 15 per cent of the literature from 2004 onwards is covered. In addition, while 75 per cent of the database is derived from original articles, 25 per cent comes from only abstracts. As a result, any information not stated in relevant abstracts might be missing from reports. It is also likely that NAPRALERT does not contain many entries on novel compounds derived from marine organisms.

ACCESS AND BENEFIT-SHARING ELEMENTS

The database does not mention ABS, nor does it comply with basic ABS principles, such as stating the source, requiring users to share benefits or monitoring subsequent use. Particularly critical is the listing of ethnomedical applications of natural products without any reference to the relevant group or community.

PubChem

PubChem is a comprehensive, public, web-based repository of the National Center for Biotechnology Information (NCBI) (Sayers *et al.*, 2011, p.D38; NCBI, 2012a; Romiti and Cooper, 2012).[1] It contains information on small chemical molecules and their biological activities provided by over 70 depositing organizations (Bolton *et al.*, 2008). It aims to support the free dissemination of biomedical information to health scientists. PubChem is organized into three distinct, but interlinked sub-databases: PubChem Substance, PubChem Compound and PubChem BioAssay. The most important one for backtracking is PubChem Compound, which contains validated chemical depiction information to describe chemical structures and comprises 27 million unique structures.

TYPE AND FORM OF STORED INFORMATION

Summary pages provide an overview of what is known about a particular compound, depending on data provided by depositors. This could include: an introductory passage, also stating the biological source species; a depiction of a chemical structure; biomedical annotations; classification and literature links; patent numbers; synonyms; chemical properties and descriptors and a list of all depositors, which have contributed information to that particular record (Box 5.4). In addition, each record links to the scientific literature via PubMed and PubMed Central.

USING THE DATABASE

Information in PubChem Compound can be accessed in various ways. The easiest is to begin at the PubChem home page and initiate a simple keyword search. This triggers an unrestricted search for the keyword in all 65 of the fields that specify compounds and may result in long hitlists of various compounds.

An alternative is to use the advanced search. This allows a keyword search restricted to one of the 65 fields, chemical properties, stereochemistry, bioassays, links to other databases, elements, depositors and other categories. Misspellings are mitigated by suggestions of correct terms.

LEGAL FORM

PubChem is hosted by the United States National Institutes of Health (NIH), which is part of the United States Department of Health and Human Services (Wang *et al.*, 2009, p.W623). NIH is the primary federal agency for conducting and supporting medical research. Under NIH, the Roadmap for Medical Research (today: NIH Common Fund) was developed, which is a series of trans-NIH programs designed to improve medical research capabilities and to increase the speed of scientific discoveries (Center for Information Technology, 2012). One of these programs is the Molecular Libraries and Imaging program. This supports nationwide collaboration to discover new chemical molecules and also maintains PubChem (Division of Program Coordination, Planning and Strategic Initiatives, 2012).

Also involved in maintaining and distributing PubChem is the NCBI, which is also part of NIH and is responsible for maintaining a variety of databases.

RULES CONCERNING FEEDING AND RETRIEVING DATA

There are two ways to feed data into PubChem. The first is through the Molecular Libraries and Imaging program. The second is uploading data over the PubChem Deposition Gateway. To upload information, users must create a deposition account, which requires information on, among others, the user's name, company or organization, job title, address and relevant online sources (NCBI, 2012c). Once the account is validated, the user has to sign a Data Transfer Agreement. Under the Agreement, the user agrees to make the data publicly accessible over PubChem without remuneration for the dissemination and exchange of scientific information related to medicine and public health.

Any person interested in the database may freely access, distribute and copy any data (NCBI, 2012d). However, it is requested that acknowledgement is given to the National Library of Medicine in any subsequent use of the data.

MONITORING OPPORTUNITIES

As it is an NCBI database, PubChem can link its entries to PubMed, the major literature database of NCBI. PubMed comprises over 20 million citations from life science journals and online books (NCBI, 2012e). It covers mainly biomedical literature from medicine, nursing, the health care system and preclinical science, but also some from non-medical fields. While PubMed does not include full text journals, it provides abstracts and links to subscription-based or free online sources of scientific papers. Providing access to abstracts alone may be a viable compromise for effective monitoring: abstracts can convey the right amount of information necessary to identify the use of a particular compound and verify the legitimacy of the use according to mutually agreed terms.

In addition, PubChem Substance provides links to the depositors of data. However, the monitoring potential of these links is rather limited, because the links often lead to other databases. Searching for actual users in these other databases often results in dead ends, because they may either require a subscription or even do not refer to any specific users.

Occasionally, PubChem refers to the trademark and the distributor. The presence of such information depends on data depositors and is not a compulsory requirement for entries.

As a new addition in 2012, PubChem increasingly refers to relevant patents. While for new compounds and products of minor market importance, the amount of patents mentioned is rather low; some entries on more successful market products might yield large amounts of patent references.

DATA LIMITATIONS

Given the sheer size of PubChem and the comprehensiveness of single records, it is difficult to clearly identify data limitations. However, the explicit focus of the database on small molecules excludes large molecules derived from marine sources, such as polymers and peptides (Apodaca, 2012). Another drawback is that depositors adopt different standards or rules for chemical structures and information. This 'noise' in data representation may confuse those users with limited knowledge of biochemical sciences. In addition, the coverage of millions of compounds makes effective maintenance of the database difficult. Finally, it is not mandatory to provide the source species, and thus finding the right molecule derived from the utilization of a particular genetic resource and linking it to its source species might be difficult.

With regard to monitoring, it is left to the discretion of data depositors as to whether they provide information on trademarks or distributors. Consequently, such information is often missing in the database. Such information may occur within the scientific literature to which PubMed refers. However, not all of the references in PubMed are freely available, so that not all relevant information can be retrieved.

ACCESS AND BENEFIT-SHARING ELEMENTS

Although ABS is not explicitly referred to, it is noteworthy that any information on the database is in the public domain. As long as appropriate acknowledgement is made, any information can be freely used and distributed.

Interestingly, under the copyright status of PubChem, NCBI mentions that some submitters of original data or the 'country of origin' of such data may claim intellectual property rights on the data. In such cases, unrestricted permission on use, copying and distribution cannot be provided. Although this provision in itself does not contribute much towards an effective implementation of ABS, it nevertheless acknowledges that certain data fall under the regulatory rights of source countries.

GenBank

GenBank is a comprehensive, freely available, regularly updated online database that contains an annotated collection of all publicly available DNA sequences for more than 300,000 organisms. In total, it contains more than 255 billion bases in over 155 thousand sequence records. As part of the International Nucleotide

Box 5.4 Elements of a record from PubChem Compound using ecteinascidin as an example (this record shows trabectedin, which is a synonym of ecteinascidin) (CID 108150). The biological source is highlighted. Since Pubchem entries are very comprehensive, only section headings including explanatory phrases (in brackets) are shown in the second half of the box Source: NCBI (2012b)

---

trabectedin - Compound Summary (CID 108150)
Also known as: Yondelis, ET-743, ecteinascidin 743, Ecteinascidine 743, Et 743, Ect 743, NSC 648766, DSSTox_CID_26880, DSSTox_RID_81984
Molecular Formula: $C_{39}H_{43}N_3O_{11}S$ **Molecular Weight:** 761.83722
InChIKey: PKVRCIRHQMSYJX-AIFWHQITSA-N

Trabectedin, also referred as ET-743 during its development, is a marine derived antitumoural agent discovered in the Caribbean tunicate *Ecteinascidia turbinata* and now produced synthetically. Trabectedin has a unique mechanism of action. It binds to the minor groove of DNA interfering with cell division and genetic transcription processes and DNA repair machinery. It is approved for use in Europe, Russia and South Korea for the treatment of advanced soft tissue sarcoma refractory to or unsuitable to receive anthracycline or ifosfamide chemotherapy.

Identification:	[Synonyms and compound information]
Related records:	[Similar compounds with annotation]
Use and manufacturing:	[Patent numbers]
Pharmacology:	[Pharmacological action]
Literature:	[PubMed citations]
Patents:	[Patent numbers and titles]
Biomolecular interactions and pathways:	[Interactions with other biomolecules]
Biological test results:	[Effects of substance on living systems]
Classification:	[Structure, use, toxicity classifications]
Chemical and physical properties:	[Weight, formulas, mass, charges, etc.]

Sequence Database Collaboration, GenBank regularly exchanges data with the European Nucleotide Sequence Database and the DNA Databank of Japan (Benson *et al.*, 2010, p.D46).

TYPE AND FORM OF STORED INFORMATION

GenBank stores its information in records. Each record includes a concise description of the sequence: the name of the genetic construct; an accession number; the scientific name of the source organism; bibliographical references; the complete genetic sequence and specific features within the sequence, such as promoter sites, coding regions for proteins and terminator sites (Box 5.5). In addition, each record contains links to entries about the taxonomic details of the source species; the protein products, described as sequences of amino acids in single-letter code and related sequences.

USING THE DATABASE

Users can access records in various ways. The most convenient approach is a keyword search on the 'Nucleotide' database on NCBI's retrieval system 'Entrez'. Keywords can be any term included within a record (sequence name, accession number, species name, etc.) or a combination thereof using Boolean operators. It is also possible to restrict or specify the search with advanced options.

The second approach is a sequence-similarity search. For this, NCBI provides the online Basic Local Alignment Search Tool (BLAST) (NCBI, 2009). BLAST is a conglomeration of programs that allows users to enter a specific query sequence of nucleotides or proteins. The program then compares the query sequence with all sequences in the database for exact or similar matches of the whole or part of the sequence. A search results in a list of selectable records that are arranged according to their degree of similarity. The purpose of BLAST is to provide scientists, who have analysed sequences, with a tool to compare their sequence with sequences from all over the world.

A third option involves downloading the database through File Transfer Protocol (FTP).

For the purpose of identifying the geographical range of source species, GenBank records provide a link to taxonomic entries, which in turn contain external links to other species information databases, including those that map distribution of species, such as the Ocean Biogeographic Information System (OBIS).

LEGAL FORM

As PubChem, GenBank is built and distributed by NCBI and thus ultimately falls under the aegis of the US government.

RULES CONCERNING FEEDING AND RETRIEVING DATA

GenBank receives its data from various sources. These include individual laboratories and large-scale sequencing projects but also the US Office of Patents and Trademarks. Data holders submit their information via direct electronic submission using various tools, depending on the type of data. These tools must first be installed and registered. By using these tools, data holders can enter sequence information directly into a form and add biological annotations, such as coding regions. The tools produce an output file that can be directly sent via email to GenBank staff or uploaded to the GenBank website. Once submitted, the data receives a quality review by GenBank staff to ensure accuracy. Although GenBank only publishes freely available data, it also allows submitters to withhold disclosure of confidential sequences until they are published. The submitters of a record maintain control of the records and can update it, if necessary.

Box 5.5 Elements of the entry on the growth hormone construct opAFP-GHc2 (AY687640), which is responsible for an increased growth rate of cultured Atlantic salmon (AquAdvantage salmon by AquaBounty). Source species are highlighted. Although this entry contains multiple indications on the source species, it only refers to two species, *(Macro-)zoarces americanus* (ocean pout) and *Oncorhynchus tshawytscha* (chinook salmon). Source: NCBI (2012f)

Synthetic construct opAFP-GHc2 growth hormone I precursor gene, complete cds

LOCUS: AY6876404061 bp   DNA linear SYN 01-AUG-2005
DEFINITION   Synthetic construct opAFP-GHc2 growth hormone I precursor
                  gene, complete cds.
ACCESSION   AY687640
VERSION   AY687640.1 GI:56691717
SOURCE   synthetic construct
ORGANISM   synthetic construct
                 other sequences; artificial sequences.
REFERENCE 1 (bases 1 to 4061)
                 AUTHORS: Hew, C.L. and Fletcher, G.L.
                 TITLE: Gene construct for production of transgenic fish
                 JOURNAL   Patent: EP 0578653-B 18-JUL-2001;
                 Seabright Corporation Limited; St. John's; Canada;
REFERENCE 2 (bases 1 to 4061)
                 AUTHORS   Agarwal-Mawal, A., Du, S.J., Hew, C.L., Yaskowiak, E.S.
                 and Fletcher, G.L.
                 TITLE: Direct Submission
                 JOURNAL: Submitted (16-JUL-2004) Aqua Bounty Canada, 20 Hallett
                 Crescent, P.O Box 21233, St. John's, NL A1A 5B2, Canada
FEATURES   Location/Qualifiers
promoter   (1..2120)
                 /note="derived from *Macrozoarces americanus* antifreeze gene"
5'UTR   (2121..2197)
                 note="derived from *Macrozoarces americanus* antifreeze gene"
CDS   (2198..2830)
                 /note="derived from *Oncorhynchus tshawytscha* growth hormone I
                 gene"
                 /codon_start=1
                 /transl_table=11
                 /product="growth hormone I precursor"
                 /protein_id="AAW22586.1"
                 /db_xref="GI:56691718"
                 /translation="MGQVFLLMPVLLVSCFLSQGAAI
                 ENQRLFNIAVSRVQHLHLLAQKMFNDFDGTLL
                 PDERRQLNKIFLLDFCNSDSIVSPVDKHETQKSSVLKLLHIS
                 FRLIESWEYPSQTLIISNSLMVRNANQISEKLS
                 DLKVGINLLITGSQDGLLSLDDNDSQQLPPYGNYYQNLGG
                 DGNVRRNYELLACFKKDMHKVETYLTVAKCR
                 KSLEANCTL"
             sig_peptide (2198..2263)
             mat_peptide (2264..2827)
                     /product="growth hormone I"
             3'UTR   (2831..2900)
                     /note="derived from *Oncorhynchus tshawytscha* growth
                     hormone I gene"
             terminator (2901..4061)
                     /note="derived from *Macrozoarces americanus* antifreeze gene"

Although the information contained within GenBank is in the public domain and can therefore be freely used and copied, appropriate acknowledgement of the National Library of Medicine must be given.

MONITORING OPPORTUNITIES

Each entry within GenBank cites the authors that submitted data on the genetic sequence and the relevant publication. In addition, GenBank provides cross-links to PubMed and PubMed Central.

DATA LIMITATIONS

As with all biological databases, GenBank is also subject to limitations in its data (Claverie and Notredame, 2007, pp.67–98). GenBank receives its data from direct depositions by scientists, sequence centres and other databases. These depositors may unintentionally submit erroneous sequencing, lineage and function data (Wesche *et al.*, 2004, p.362; Bidartondo, 2008, p.1616). Moreover, depositors may use varying data formats and schemes and apply different nomenclature (Sumithiradevi and Punithavalli, 2009, p.141). This could lead to incomplete, inaccurate and even ambiguous entries, data duplication or inconsistent entries (Apiletti *et al.*, 2006, pp.1–2). These limitations are particularly relevant to scientists in the field of molecular biology who require highly accurate data. However, these limitations should not impair the ability to link the product to its source species, because the species is always stated and unambiguously identified through genetic techniques.

Monitoring opportunities are sparse, because GenBank does not refer to any patents, brands or firms marketing the brand. Papers mentioned by PubMed may contain more information in this regard, if they are freely accessible.

ACCESS AND BENEFIT-SHARING ELEMENTS

GenBank completely fails to conform to some of the basic principles of ABS. For example, the source country from which the genetic sequence was obtained is not mentioned. It is unclear whether researchers have complied with prior informed consent and mutually agreed terms. Data can be used for any purpose; a change of intent by third party use is not regulated.

National Center for Biotechnology Information Protein database

The protein database is the third NCBI database to be examined within this work. The protein database contains amino acid sequences created from the translation of coding regions provided in GenBank, Swiss-Prot and other databases (NCBI, 2012g). Due to the similarity to PubChem and GenBank, this database will only be introduced briefly.

TYPE AND FORM OF STORED INFORMATION

The records held by the protein database are very similar to GenBank records. They include: a concise description of the protein; an accession number; the Latin name of the source species and its taxonomy; references to the scientific literature; links to the relevant genetic sequence encoding for the protein; features of the protein sequence, such as functional regions, catalytic sites and substrate binding sites and the protein sequence itself, represented in single-letter amino acid code.

USING THE DATABASE

As with GenBank, the easiest approach for finding information in the protein database is a simple keyword search yielding a hit list of exact and similar matches. An advanced search by using limits, specified fields and Boolean operators is also possible.

LEGAL FORM

Because it is part of the NCBI assembly of databases, the protein database also belongs to the public domain and falls under the ultimate authority of the US government.

RULES CONCERNING FEEDING AND RETRIEVING DATA

The protein database receives its information primarily from other databases, which in turn receive their information from individual scientists, research institutes, single projects involved in protein sequencing and other databases.

Users who wish to retrieve and use data may do so freely, the only condition being proper acknowledgement of the National Library of Medicine.

MONITORING OPPORTUNITIES

As with GenBank, the protein database refers only to the authors that submitted data on the protein, citing their publication. In addition, the database provides links to research papers on PubMed and PubMed Central, if freely accessible research papers exist.

DATA LIMITATIONS

Due to its close relationship to GenBank, the same data limitations apply in principle for NCBI's protein database.

ACCESS AND BENEFIT-SHARING ELEMENTS

The ABS deficiencies of this database are the same as for all above databases and are therefore not reiterated here.

Universal Protein Resource

The Universal Protein Resource (UniProt) is a publicly accessible, online repository of protein sequences and relevant functional annotations (Jain *et al.*, 2009; UniProt Consortium, 2010, 2012a). It strives to be the central repository on proteins and to provide all known relevant information on any particular protein. It has four components: the UniProt Knowledgebase, the UniProt Archive, the UniProt Reference Clusters and the UniProt Metagenomic and Environmental

Sequences database. Of these, UniProt Knowledgebase is the central access point for integrated protein information on 524,000 reviewed (Swiss Institute of Bioinformatics, 2012b) and over 13 million unreviewed (European Bioinformatics Institute, 2012) protein sequences. It will be the focus of this section.

TYPE AND FORM OF STORED INFORMATION

The entries on protein sequences take the form of datasheets, which contain large amounts of information. This includes: names and biological origin; protein attributes; general annotation; ontologies (keywords and gene ontology); sequence annotations (features of specific protein parts); the relevant sequence in single-letter code; references to the published scientific work that has submitted protein information; cross-references to GenBank and other external data collections; entry information (accession number, history of updates and modifications) and relevant documents (Box 5.6). Much of the information in these categories is linked to other websites containing further information.

USING THE DATABASE

The primary access point to the data and documentation of UniProt is the online interface. The interface provides various tools for querying data. The primary tools are full-text and field-based text searches using keywords. Additional tools include sequence similarity searches by using BLAST, multiple sequence alignment and batch retrieval (accessing multiple entries by using unique identifiers, e.g., accession numbers). UniProt is designed to allow successful searches of most data through the full text search without requiring detailed knowledge of data or search syntax. Once an entry is opened, the information provided can be selected to retrieve further information. For example, selecting the source species yields detailed taxonomic information and additional Latin or common names. A lot of the other information can also be selected, such as keywords that lead to background information on biological processes, reference links that provide relevant scientific articles and links to additional databases with further services.

LEGAL FORM

UniProt was created and is maintained by the UniProt Consortium, which comprises the European Bioinformatics Institute, the Swiss Institute of Bioinformatics and the Protein Information Resource. The UniProt Consortium has decided to apply the Creative Commons Attribution-NoDerivs Licence to all copyrightable parts of the database (UniProt Consortium, 2012b). Thus, the database and any of its elements are in the public domain.

RULES OF FEEDING AND RETRIEVING DATA

As well as receiving data from collaborating databases, UniProt allows for the individual to feed in protein information. For feeding data, submission occurs through a series of online forms, where contact details, protein names, source organisms, citation of the scientific work, properties of the protein, the sequence

Box 5.6 Elements of a UniProt entry using Deep Vent DNA Polymerase as an example (Q51334). The source species (*Pyrococcus* sp) is highlighted. Portions of the entry are omitted for reasons of manageability. Source: UniProt Consortium (2012a)

Q51334 (DPOL_PYRSD) Reviewed, UniProtKB/Swiss-Prot
Names and origin

       Protein names

              Recommended name: DNA polymerase EC=2.7.7.7

              Alternative name(s):Deep Vent DNA polymerase

       Gene names: pol

       Organism: *Pyrococcus* sp. (strain GB-D)

       Taxonomic identifier: 69013 [NCBI]

       Taxonomic lineage: Archaea > Thermococci > Thermococcales > Thermococcaceae > Pyrococcus

Protein attributes	[sequence length, status and processing]
General annotation	[Function, catalytic activity, biotechnological use, etc.]
Ontologies	[Keywords on biological processes and molecular functions]
Sequence annotation	[Features of the enzyme and their location on the amino-acid chain]
Sequences	[The full sequence in single-letter amino-acid code]
References	[PubMed citations]
Cross-references	[Links to other database entries]
Entry information	[Entry name, history and status and accession number]

---

itself and confidentiality information is asked for in successive steps. With regard to confidentiality, data contributors must be aware that the information belongs to the public domain once it has been published in the database. UniProt does allow data to be kept confidential until publication of the data in a scientific journal or until after a defined date, set by the data contributor.

Because the database and its elements belong to the public domain, users are free to copy, distribute, display and make commercial use of the database in all countries, provided appropriate credit is given. For any kind of alteration, transformation or modification of the database, permission must be obtained.

MONITORING OPPORTUNITIES

As with the above-mentioned databases, UniProt contains a reference list on research papers and abstracts. As already stated, abstracts may be the right tool for monitoring compliance. They state the use of a particular compound in a concise and generally understandable way. Thus, providers can quickly assess how a particular compound has been used.

A cross-reference section supplements the reference list. The cross-reference section provides external links to other databases that contain information on the same protein. This could include databases on the protein sequence, source organism and protein family and domain. Each of these databases also includes a reference list or a list of data depositors.

DATA LIMITATIONS

Although UniProt also relies mainly on data submission from science, the error rate of entries is much lower than of other databases, because it is manually curated (Schnoes *et al.*, 2009, p.1). However, only a minor part of the whole database is reviewed, which leaves the major part unreviewed and error-prone.

With regard to monitoring, the database occasionally states the trademark and relevant firm, but does not make this a mandatory requirement for its entries. In addition, the database only mentions central research papers.

ACCESS AND BENEFIT-SHARING ELEMENTS

UniProt does not deal with ABS. Source countries are not acknowledged, and there is no obligation for benefit sharing and any subsequent utilization of information is not monitored. The only, though very weak, monitoring mechanism is the obligation to credit UniProt when using information for both commercial and non-commercial purposes.

Linking the source species to its geographical distribution

This section analyses the databases that describe the geographical ranges of marine species across the globe. These databases are analysed for the same criteria as above, except for monitoring opportunities, because generating data for such databases does not qualify as a utilization of genetic resources.

Ocean Biogeographic Information System

The Ocean Biogeographic Information System (OBIS) is a web-based, open-access gateway to many datasets with a primary aim of providing taxonomically and geographically resolved data on locations of marine life in the ocean environment (Grassle and Stocks, 1999; Grassle, 2000; IOC, 2012). This data comes from credible and authoritative sources, including individual scientists, government agencies, museums, universities, commercial companies and non-governmental organizations. To date, OBIS has more than 30 million location records of 126,000 species from 898 interoperable databases (OBIS, 2011).[2]

TYPE AND FORM OF STORED INFORMATION

The primary data held by OBIS is geo-referenced records of particular species (or higher taxonomic groups), recorded as discrete geographical coordinates on the globe. This is represented either as abundance per unit grid or as location points

(Figure 5.5). In addition to geographic data, OBIS provides metadata on species information and data sources for each location point.

Species information includes: the Latin species name and common names; the position of the species in taxonomic lineage; its geographic coverage; temporal coverage of collection or observation of specimens and environmental information on depth range, temperature and various chemical properties. In addition, OBIS links to other collaborating information systems for marine species (Encyclopedia of Life, Catalog of Life, World Register of Marine Species, Integrated Taxonomic Information System, Barcode of Life, GenBank, Google, uBio and FAO) and mapping systems (Global Biodiversity Information Facility, DiscoverLife, AquaMaps and the Kansas Geological Survey Mapper).

Metadata on records are made available in basic summaries, which describe features of individual records. Features include, among others: taxonomic, geographic, temporal and habitat coverage; collection method; data source, scientific and technical contacts and approximately 70 additional features, if the information is available.

USING THE DATABASE

The user accesses any data on marine species within OBIS over an online interface. The primary means to retrieve data from the interface is to perform a simple species name search for the Latin name or common name via the search box. Once the species is selected, the interface produces several new windows with species information, maps of global occurrences and relevant sources. The map display can be adjusted and overlaid with, for example, EEZs, Large Marine Ecosystems and Marine Ecoregions of the World. From the window listing data sources, the user can download the complete data, including latitude and longitude data, in various formats, which can be further analysed in geographical information system (GIS) programs. To link the source species to source countries, this would mean overlaying occurrences with EEZs.

In addition to the simple name search, a user can also perform an advanced search taking into consideration specific datasets, user-defined regions and various oceanographic parameters.

LEGAL FORM

From 2000–2010, the Census of Marine Life programme operated OBIS with funds from the Alfred P. Sloan Foundation.[3] After the conclusion of the Census, OBIS was fully integrated into the International Oceanographic Data and Information Exchange (IODE) Programme of the International Oceanographic Commission (IOC) of the United Nations Educational, Scientific and Cultural Organization (UNESCO) (IOC, 2009, para12(i)).

RULES CONCERNING FEEDING AND RETRIEVING DATA

As indicated above, any organization, consortium, project or individual can contribute to OBIS, as long as the source is credible and authoritative (OBIS, 2011).

However, submitted data must still pass through a series of technical and quality controls before it is published by OBIS. This process ensures scientific correctness and authority.

There are two possible ways to feed data into the OBIS system. Either the provider hosts data, or OBIS does. In the first case, the provider keeps the data locally on a server running on special software, which is freely distributed by OBIS and that can respond to queries from the OBIS system. If OBIS hosts the data, then the provider needs to submit the relevant dataset in electronic format to an OBIS-supporting organization, the central data portal or another OBIS-recognized data provider.

Because OBIS is committed to keeping its data free and openly accessible, data contributors are not remunerated for their data. This could be problematic for data holders with sensitive data on commercial or rare species. As a compromise, OBIS offers to publish the data with a lower spatial resolution or to give a bounding box instead of a point location.

For the retrieval of data, OBIS prescribes three rules to which users of the data must agree. First, users must cite the data source according to a model citation scheme. Second, for information purposes, users must submit the full citation of any printed or electronic publication made that cites OBIS or constituent parts. Finally, users must acknowledge the limitation of data in OBIS.

DATA LIMITATIONS

Despite its comprehensiveness and the high quality of its data, OBIS still suffers from various drawbacks owing to our incomplete knowledge of the oceans (OBIS, 2011). First, some oceanic realms and taxonomic groups are better surveyed than others: coastal areas are better explored than the deep sea, while the oceans of the northern hemisphere are better sampled than those of the southern hemisphere, and the distribution of marine vertebrates is more reliably represented than that of invertebrates. Least well documented are microbial species. Second, most marine species have not been recognized or named. Third, many marine species that have been described as a single species are actually several species, while other multiple species should be combined into a single species. Fourth, many species are only represented by very few locations, and so it is not possible to infer a valid full distribution range. Fifth, many databases are not yet connected to the OBIS system. As a result, much of the knowledge about species occurrences is not yet available over the OBIS system. Sixth, some species' records derive from decades-old collections or observations. This could cause problems with far-reaching consequences in situations where a population of a specific species becomes extinct in one country but a population of same species survives in a neighbouring country. If the extinction is not recognized because old records are not updated, both countries would qualify as source countries and could therefore claim benefit sharing. This is a very unfair situation, because one country will be rewarded despite the fact that it has failed to maintain the survival of the species. Finally, no matter how numerous records of a particular species are, one might argue that mapping discrete occurrences can only ever approximate true species distributions. This would imply that all the source countries for a specific product from marine genetic resources cannot be identified, particularly when there are only a few records for one species.

Despite these limitations and the high degree of scientific effort required to ameliorate these deficiencies, OBIS still constitutes the most authoritative source on marine species distribution. The quality of OBIS data is rapidly improving (OBIS, 2011). Moreover, it provides an incentive for states to better explore their marine biodiversity in order to increase their chances of qualifying as a source state once a product from a particular species from their waters is developed. It would be unwise to reject this approach of identifying source countries of products from marine genetic resources.

KANSAS GEOLOGICAL SURVEY (KGS) MAPPER

The KGSMapper is a web-based biogeographic tool which complements OBIS data. It calculates the values of environmental parameters where a species has been recorded and then displays all other places globally where values are statistically similar. It can therefore model the potential range of the species (Guinotte *et al.*, 2006; KGS, 2012). Data can be fed into the database via a direct link from species pages within OBIS. A great strength of the KGSMapper is that it can produce accurate range maps from as few as 20 location records within OBIS (Costello, 2006). Thus, by identifying all area where a particular species might occur, the KGSMapper eliminates the weakness of inferring a species' range only from known location points.. Consequently, countries that have no location record within OBIS could still qualify as a source country, if they take part in the modelled area of the KGSMapper.

However, its great strength is also the great weakness of the KGSMapper in ABS contexts. Many species within OBIS are represented by less than 20 records. For example, the source species of the multi-million dollar drug Yondelis, the Caribbean mangrove crab *Ecteinascidia turbinata*, yields only eight location records. In such cases, it is not only difficult to justify the modelled geographic range scientifically, but it also suggests that the general applicability for identifying source countries should be scrutinized. In other words, it is unfair to include as source countries all those countries which have no location records but that fall into modelled ranges of species with 20 or more records, but to then exclude countries from being source countries when species are represented by less than 20 records. It must be concluded that this approach is not suitable for linking a source species to its geographical range in order identify states eligible for benefit sharing.

ACCESS AND BENEFIT-SHARING ELEMENTS

OBIS does not explicitly recognize ABS and is not directly involved in the utilization of genetic resources. However, it represents an exemplary approach for non-commercial research when engaging in ABS processes because it benefits all parties involved and promotes the objectives of the CBD concerning conservation and sustainable use.

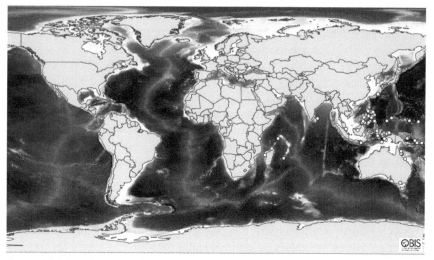

Figure 5.5 Map of an exemplary species distribution in OBIS, using the cone snail, *Conus magus*, as an example. The map shows discrete location points

Source: IOC (2012) and relevant datasources[4]

Data providers retain ownership of their data and have a standard platform to publish their data, thus maximizing their scientific recognition. Source countries have a tool to meet their reporting obligations to the CBD, are acknowledged by the location of a record and benefit in various other ways, as reflected by the Nagoya Protocol.[5] The global scientific community benefits from an increased availability of scientific data, which can be used for testing hypotheses on marine biodiversity, marine ecology and evolutionary processes (Grassle and Stocks, 1999, p.13). OBIS may also contribute to the management of marine biodiversity and ecosystems by providing the necessary information to, for example, delineate marine protected areas and other specific sea areas.

With regard to intellectual property rights, OBIS does not claim ownership or rights on the datasets it publishes. All rights remain with the data source. In addition, all data published are labelled with the organization and database from which they came, and a standard citation is provided.

Global Biodiversity Information Facility

The Global Biodiversity Information Facility (GBIF) is an international government-initiated and government-funded organization focusing on publicly disclosing global biodiversity data through their web portal (GBIF, 2007; Lane and Edwards, 2007; GBIF, 2012a). The portal allows information to be accessed on two basic types of data: species occurrence records and classifications of organisms in the taxonomic hierarchy. Currently, GBIF contains more than 267 million occurrence records from over 315 data publishers and over 10,000 data sources. The main data publisher for the marine environment is OBIS.

This could render a closer examination futile; however, a full analysis of GBIF is necessary because it: a) contains additional data on marine organisms from a plethora of minor data publishers; b) provides location information of organisms without specific coordinates and c) prescribes rules of conduct relevant to ABS.

TYPE AND FORM OF STORED INFORMATION

The GBIF portal provides overview pages of species, countries and datasets. Key to these three types of overview pages are interactive global maps showing the distribution of geo-referenced occurrence records for a species or group of species (Figure 5.6). These maps only show records with coordinates on a colour-coded 1 degree grid indicating the abundance of occurrences; records without coordinates are not mapped but their source country is still included in the database.

The overview pages for species summarize specific data and allow a more detailed exploration of the data. Specific data is sub-divided in various areas of the overview page. The top provides the Latin species name together with its taxonomic lineage. An 'action box' contains quick links to details of each occurrence record of a particular species, countries with occurrences (Box 5.7), datasets with occurrences and data download options. The interactive map allows the visualization of occurrences that have coordinates. Below the map, the relevant datasets that provided geo-referenced occurrence records are listed.

The overview pages for countries summarize the available information on all species that occur within the marine and terrestrial boundaries of a particular country. The layout of the country overview page is similar to the species overview page.

A database overview page provides information about the institutions, organizations and datasets involved in providing the occurrence records and visualizing geo-referenced occurrences of the dataset through the map.

As already indicated, all data on the overview pages is predicated upon occurrence records. Each record provided is given a unique identification number and contains a multitude of information on the relevant dataset (data publishers, dataset name and rights, institution codes, collector names, data of collection, images), the taxonomy of the recorded species and geographic attributes (country, coordinates, etc.).

USING THE DATABASE

There are two basic ways of finding information on species, countries or datasets. Users can perform direct searches by putting keywords into a search box. This triggers a search for results, which are then grouped into four categories, namely 'scientific names', 'common names', 'countries' and 'datasets', from which the user may choose the appropriate entry. Alternatively, it is possible to select a species, country or dataset by browsing alphabetical lists. Regardless of which type of search users conduct, they first have to agree to the GBIF Data Use Agreement.

LEGAL FORM

The memorandum of understanding lays down the legal form of GBIF (GBIF Secretariat, 2010). Under the memorandum, GBIF is an open-ended international coordinating body which aims to 'promote, co-ordinate, design, enable and implement the compilation, linking, standardization, digitization and global dissemination and use of the world's biodiversity data' (paragraphs 2.1 and 3.1). GBIF comprises a governing board (countries, economies, international organizations) (4), a secretariat (6) and an executive secretary (7). Like OBIS, GBIF is an open-access facility promoting equal access and the free dissemination of biodiversity data (8.2–8.3).

RULES OF FEEDING AND RETRIEVING DATA

Data owners that want to publish their data have two main ways to do so. The first involves the maintenance of a server with specialized software installed, registration through a GBIF registration page (GBIF, 2012b) and signing of the GBIF Data Sharing Agreement (GBIF, 2012c). Under the Agreement, data publishers agree, among others, that: a) their biodiversity data are openly and universally available; b) GBIF asserts no intellectual property rights on the data; c) the necessary agreements with the original owners of the data have been made; d) their data are accurate and e) they take responsibility regarding access restrictions to sensitive data on, for example, endangered species (paragraphs 1.1–1.5). The second way involves submitting data directly, which also requires signing the Agreement.

Data users must agree to the GBIF Data Use Agreement before they can access any information over the portal (GBIF, 2012d). Under that Agreement, users: a) acknowledge that the quality and completeness of data cannot be guaranteed; b) agree to respect restrictions to sensitive data; c) make attribution of use and d) publicly acknowledge the data publishers whose biodiversity data they have used (1.1–1.4). For the latter point, the Agreement provides a standard format to cite data retrieved from the GBIF network (2).

DATA LIMITATIONS

Despite the outstanding quality of GBIF, the database repeats the same data deficiencies for the marine environment as OBIS, because the latter is the main data provider of information on marine organisms. In addition to those shared deficiencies, GBIF itself highlights potential issues with names and taxonomic classifications of organisms (GBIF, 2007, pp.2–3). Depending on the taxonomic classification used, scientific names for the same organism may differ. Although GBIF has implemented some strategies to improve this situation, many records remain uncorrected. Another problem is that sampling efforts are usually limited, scattered and not standardized, and inventories are biased towards easily accessible sampling sites. Therefore, biodiversity data are still scarce, biased and sometimes of poor quality (Hortal *et al.*, 2007, p.853). In addition, species records may be outdated and a recognition of geographic and environmental variations that affect distribution of organisms is often lacking. Despite the validity of this criticism, the approach of GBIF to integrate information from all sources with an evolutionary improvement of the databases, makes it the best instrument to show the global distribution of biodiversity.

ACCESS AND BENEFIT-SHARING ELEMENTS

In various places, GBIF gives some indication as to how it conforms to ABS. For example, under paragraph 1.3 of the GBIF Data Sharing Agreement, data publishers are responsible for making all the necessary agreements with the original owners of the data before they can make data available through the GBIF network. This is an oblique reference to the establishment of prior informed consent and mutually agreed terms. A more explicit reference occurs under the memorandum of understanding. Under paragraph 8.7 on 'legitimacy of data collection', GBIF asks for 'reasonable assurances from the data publisher/holder that such access was consistent with applicable laws, regulations and any relevant requirements for *prior informed consent*' (emphasis added). These provisions constitute guidance, albeit weak, for complying with ABS provisions. The memorandum and the Agreement say nothing about cases of non-compliance. If such a case were to be discovered, it is likely that the relevant data would be withheld from disclosure or removed from the network. The provisions of the memorandum are further weakened by its non-binding nature (preamble).

Even worse is the almost complete disregard of the subsequent utilization of biodiversity data by third parties. The only mention in the memorandum obliges GBIF not to limit further 'non-commercial use and dissemination of [biodiversity]

Figure 5.6 Map of a GBIF record, using *Conus magus* as an example. Please note that the map only shows occurrences with coordinates (584 of 938 total occurrences). All occurrences per country (also without coordinates) can be accessed through the quick link 'countries with occurrences' in the action box

Source: Biodiversity occurrence data accessed through GBIF Data Portal, data.gbif.org, 16 November 2012. The underlying data of the maps come from multiple data providers[6]

Box 5.7 GBIF countries with occurrence records of *C. magus*. The occurrences of each country can be selected for a detailed analysis of each occurrence record. Source: Biodiversity occurrence data accessed through GBIF Data Portal, data.gbif.org, 16 November 2012. The underlying data of the maps come from multiple data providers

---

Occurrence search – countries with occurrences
Australia
Central African Republic
Comoros
Fiji
France
French Polynesia
India
Indonesia
Japan
Kiribati
Madagascar
Malaysia
Marshall Islands
Mauritius
Mexico
Micronesia, Federated States of
Mozambique
New Caledonia
Oman
Palau
Philippines
Seychelles
Singapore
Solomon Islands
South Africa
Sri Lanka
Tanzania, United Republic of
United States
Vanuatu
Vietnam
Yemen

---

data, apart from due attribution of their source' (3). There is no guidance for commercial uses, no obligation to share benefits, and no monitoring of subsequent commercial use. In order to remain in compliance with the approach taken by the GBIF Data Sharing Agreement, users with commercial intentions could be obliged under the GBIF Data Use Agreement to warrant that they have made the necessary agreements with the original owners of the data that they can utilize the data for commercial purposes.

With regard to intellectual property rights, GBIF does not assert any rights on the data. Data providers retain all rights and responsibilities associated with the data they make available (8).

Exemplary application of selected databases

This section applies the biological databases introduced above (except OBIS) to a sample set of products from marine genetic resources (biological molecules) in order to identify the source states as well as relevant users and uses. It is important to note at this point that some databases are subject to rapid development. For example, PubChem introduced extensive patent references in entries in 2012, and public databases in general are complemented constantly with the latest scientific findings. As such, the detailed information might be outdated; nevertheless, the following results should provide an overview on what databases are capable of achieving in ABS contexts.

Tracing products from marine genetic resources to their source countries

This section illustrates how selected databases can be applied in two steps in order to track biological molecules back to their source countries. Products introduced in Table 2.1 (natural compounds and commercial derivatives) were used as a sample set to test the databases.

In the first step, databases of biological molecules were analysed to see whether they mentioned the source species of a molecule. For nucleic acid, GenBank was used, for proteins, the NCBI protein database and UniProt were used and for all other types of molecules, the other databases were used (DMNP, RÖMPP, Merck Index, NAPRALERT and PubChem). To find the relevant entry in the database, simple keyword searches for the names of the biological molecules were conducted. Keywords included names and synonyms derived from the scientific literature, on both the natural compound originally isolated and on commercial derivatives. Where simple keyword searches proved unsuccessful, additional, more technical criteria were used. These included CAS number, molecular formula, molecular structure, IUPAC nomenclature (a worldwide applied nomenclature of chemical compounds developed by the International Union of Pure and Applied Chemistry), SMILES ('simplified molecular input line entry specification', which is a specification unambiguously describing structures of molecules using ASCII strings) and weight, all derived from the scientific literature. In cases where a search criterion provided a product entry (a hit), the entry was analysed for the biological source of the compound. The results are listed in Tables 5.3 and 5.4.

These tables contain a large amount of information. The left column gives the name (or type) of the natural compound in bold. Directly below each natural compound, the names of market products and their natural derivative are listed. The header row gives the different databases in bold. Each database column is

Table 5.2 Summary of the databases analysed for their content, application, legal form, rules of feeding and retrieving data, data limitations and ABS elements. These databases will be applied to trace products from utilization of marine genetic resources back to their source countries

Database	Type of information	Number of compounds/ records	Using the database	Legal form	Charge (US$)	Rules	Monitoring	Limitations	ABS elements
Dictionary of Marine Natural Products	Biochemicals	35,000	Book, CD-ROM, online (browser plug-in)	Private	Yes (595/year)	Registration, acknowledgement (user)	Research papers	Some limitations on specific biochemical classes	n/a
RÖMPP Online	Biochemicals	60,000 (313 marine)	Book, CD-ROM, online	Private	Yes (on inquiry)	Licence agreement (user)	Research papers	Few marine compounds	na
Merck Index	Biochemicals	10,000 (marine unclear)	Book, CD-ROM, online	Private	Yes (academic: 190, business: 490)	Licence agreement (user)	Research papers Patents Trademarks Manufacturers	Marine coverage unclear	n/a
NAPRALERT	Biochemicals	n/a	Online, download	Private	Yes (0.25–0.5/ citation)	Registration	Research papers (Patents)	Fragmented coverage 2004–present	n/a
PubChem	Biochemicals	27 million (marine unclear)	Online, download	Public	No	Data Transfer Agreement (provider), acknowledgement (user)	Research papers Data depositor Patents (trademarks)	Only small molecules, maintenance difficult	Raises awareness of country of origin
GenBank	Nucleic acids	155,000 sequences (marine unclear)	Online, download	Public	No	Registration (provider), acknowledgement (user)	Research papers	Maintenance difficult, data errors	n/a
NCBI Protein	Proteins	n/a	Online, download	Public	No	Acknowledgement (user)	Research papers	Maintenance difficult, data errors	n/a

Database	Type of information	Number of compounds/records	Using the database	Legal form	Charge (US$)	Rules	Monitoring	Limitations	ABS elements
UniProt	Proteins	524,000; unreviewed: 13 million	Online, download	Public	No	Registration (provider), credit (user)	Research papers Cross-references to other databases	Maintenance difficult, data errors	n/a
OBIS	Species distributions (marine)	30 million	Online, download	Public	No	Quality control (provider); acknowledge source, OBIS, and limitations	n/a	Scientific uncertainty, certain biases, incompleteness	n/a
GBIF	Species distributions (terrestrial and marine)	267 million	Online, download, CD-ROM	Public	No	Data Sharing Agreement (provider), Data Use Agreement (user)	n/a	Taxonomic classification, incompleteness	Provider-source agreements, PIC assurance

Table 5.3 Linking biological molecules to their source species using selected databases. DMNP: Dictionary of Marine Natural Products, NAPRALERT: Natural Products Alert, NCBI: National Center for Biotechnology Information. 'Y' under 'Hit' indicates presence of a product entry within a database. Where an entry mentions the source species, it is listed in Latin species name. A '-' indicates that the search did not yield a hit or the database does not apply to this type of molecule. For certain classes of compounds (halogenated furanones, DNA polymerases and superoxide dismutases) it was not possible to retrieve the name for the single natural compound responsible for a market product. The brand Spirulina Pacifica is sold for a range of compounds (vitamins, proteins, etc.) which are too numerous to list on their own. Concerning 'Cefixime', A. chrysogenum is synonymous to the more commonly applied C. acremonium

Natural compound / Market product (derivative)	DMNP		RÖMPP		Merck Index		NAPRALERT		PubChem		NCBI protein		UniProt	
	Hit	Species	Hit	Species	Hit	Species	Hit	Species	Hit	Species	Hit	Species	Hit	Species
Nereistoxin	Y	Lumbriconereis heteropoda	Y	L. heteropoda	-	-	Y	-	Y	L. heteropoda	-	-	-	-
Evisect S (Thiocyclam)	-	-	Y	-	-	-	-	-	Y	-	-	-	-	-
Bancol (Bensultap)	-	-	Y	L. heteropoda	-	-	-	-	Y	-	-	-	-	-
Padan (Cartap)	-	-	Y	L. heteropoda	Y	L. heteropoda	-	-	Y	-	-	-	-	-
Halogenated furanones	Y	Delisea pulchra	(Y)	-	-	-	Y	D. pulchra	Y	-	-	-	-	-
Pearlsafe/Netsafe	-	-	Y	-	-	-	-	-	-	-	-	-	-	-
Dichloroctylisothiazolon	-	-	Y	-	Y	-	-	-	Y	-	-	-	-	-
Sea-Nine 211	-	-	Y	-	Y	-	-	-	-	-	-	-	-	-
Rhamnolipid-2	Y	Pseudomonas aeruginosa	Y	-	-	-	Y	P. aeruginosa	Y	P. aeruginosa	-	-	-	-
JBR425	-	-	-	-	-	-	-	-	-	-	-	-	-	-
DNA polymerases	-	-	-	-	-	-	-	-	-	-	Y	480K hits	Y	350K hits
VentR DNA Polymerase	-	-	-	-	-	-	-	-	-	-	Y	Thermococcus litoralis	Y	T. litoralis
Therminator III Polymerase	-	-	-	-	-	-	-	-	-	-	-	-	-	-
Deep VentR DNA Polymerase	-	-	-	-	-	-	-	-	-	-	Y	Pyrococcus sp	Y	Pyrococcus sp
Superoxide dismutases [Mn]	-	-	(Y)	Thermus thermophilus	Y	All organisms	Y	Only terrestrial	-	-	Y	+8K hits (also: T. thermophilus)	Y	+3K hits (also: T. thermophilus)
Venuceane	-	-	-	-	-	-	-	-	-	-	-	-	-	-
Spirulina Pacifica	-	-	-	-	-	-	-	-	-	-	-	-	-	-

Natural compound / Market product (derivative)	DMNP Hit	DMNP Species	RÖMPP Hit	RÖMPP Species	Merck Index Hit	Merck Index Species	NAPRALERT Hit	NAPRALERT Species	PubChem Hit	PubChem Species	NCBI protein Hit	NCBI protein Species	UniProt Hit	UniProt Species
Astaxanthin	Y	Haematococcus pluvialis	Y	-	Y	-	Y	H. pluvialis (+ 56 additional species)	Y	-	-	-	-	-
BioAstin	-		-		-		-		-		-		-	
Docosahexaenoic acid	Y	Microciona prolifera	Y	(Y: algae)	Y	Sardina ocellata	Y	Chlorella hirataii	Y	-	-	-	-	-
life'sDHA/DHASCO	-		Y	-	-		-		-		-		-	
Pseudopterosins	Y	Pseudopterogorgia elisabethae	Y	P. elisabethae	-		-		Y	P. elisabethae	-		-	
Resilience (Pseudopterosin A)	-		-		-		Y	P. elisabethae	Y	P. elisabethae	-		-	
Spongothymidine	Y	Cryptotethia crypta	Y	-	-		Y	C. crypta	Y	-	-		-	
Spongouridine	Y	C. crypta, E. cavolini	Y	-	-		Y	C. crypta, E. cavolini	Y	-	-		-	
Zovirax (Acyclovir)	-		Y	-	Y	-	-		Y	-	-		-	
Combivir (Zidovudine)	-		Y	-	Y	-	-		Y	-	-		-	
Trizivir (Zidovudine)	-		Y	-	Y	-	-		Y	-	-		-	
Cytosar-U (Ara-C/Cytarabine)	-		Y	-	Y	-	-		Y	-	-		-	
Arasena-A (Vidarabine/Ara-A)	Y	E. cavolini, Streptomyces sp	Y	Streptomyces sp	Y	Streptomyces antibioticus	Y	E. cavolini, S. antibioticus	Y	S. antibioticus	-		-	
ω-conotoxin M VIIA	Y	Conus magus	Y	C. magus	Y	C. magus	Y	C. magus	Y	C. magus	Y	C. magus	Y	C. magus
Prialt (Ziconotide)	Y	C. magus	Y	C. magus	Y	C. magus	-	-	Y	C. magus	Y	C. magus	Y	C. magus
Ecteinascidin-743	Y	Ecteinascidia turbinata	Y	E. turbinata	Y	E. turbinata	Y	E. turbinata	Y	E. turbinata	-		-	
Yondelis (Trabectedin)	Y	E. turbinata	-		Y	E. turbinata	-		Y	E. turbinata	-		-	
Cefixime	-		Y	Acremonium chrysogenum	Y	-	Y	Cephalosporium acremonium	Y	-	-		-	
Surpax	-		-		Y	-	-		Y	-	-		-	

Table 5.4 Linking gene products to their source species using the database GenBank. 'Y' under 'Hit' indicates presence of a gene entry. Source species are listed by Latin species names

Gene (construct)	GenBank	
Market product		
	Hit	Species
opAFP-GHc	Y	*Zoarces americanus, Oncorhynchus tshawytscha*
AquAdvantage salmon	Y	*Z. americanus, O. tshawytscha*
Green fluorescent protein gene	Y	*Aequorea victoria* (8 hits) (other marine: *A. macrodactyla, Montastraea faveolata*)
GloFish	-	-
Red fluorescent protein gene	Y	*Discosoma* sp (6 hits)
GloFish	-	-

subdivided into two. The first part ('hit') indicates whether the database contains an entry on a natural compound or a commercial product by 'Y' or '-'. If 'Y' is in parentheses, the database does only contain an implicit reference to the molecule within general entries on the compound class, e.g., halogenated furanones. The second part ('species') indicates whether the entry contains a reference to the biological source (in Latin species name). Sometimes the biological source was indicated broadly, e.g., terrestrial organisms. Once the source species had been identified, it was used to infer the source states in the second step.

In the second step, the GBIF geographical database was applied to link the source species to its source states. To do this, Latin species names inferred from the first step, as described above, were used as search terms in the search box of the GBIF online portal. This produced species overview pages. By selecting the command 'countries with occurrences' within the action box of a species overview page, the source states where a particular species was collected or observed were retrieved. These source states are listed in Table 5.5.

The table only lists source states for the biological molecules for which databases contained entries which referred to the biological source. All other molecules, i.e., those without database entries, were omitted. For convenience, the left-hand column refers to only one type of molecule (natural compound or commercial product) even if databases had entries on both. Where databases mentioned different source species for the same molecule, such as for Docosahexaenoic acid, the source states are subdivided in the table according to source species. The right-hand column acknowledges that not every record is accompanied by a reference of a source state. It therefore shows the ratio between records which give source states and those which do not. This provides an indication as to how many more source states may actually exist.

Monitoring users of marine genetic resources

This section examines how selected databases can be applied in order to monitor both users and uses of genetic resources. To do this, the biological molecules introduced in Table 2.1 were again used as a sample set with which to test the databases. Each database which had an entry for a biological molecule (see Tables 5.3 and 5.4) was analysed for the number of research papers, patents, commercial brands, marketing firms or cross-references it contained. The results are listed in Table 5.6.

If all databases referred to only one type of 'monitoring instrument', e.g., research papers, the table also keeps to that type of instrument. Where a database mentioned instruments not given by other databases (e.g., DMNP mentions a brand name, while RÖMPP does not), it is indicated for this database, while the entries of other databases without such a reference are marked '-'.

It is important to bear several issues in mind. First, the table mirrors the contents of the databases only and is therefore not exhaustive. Additional information could be found by looking at the research papers listed in the databases. Given the huge amount of literature, this analysis is not conducted here.

Second, PubChem, GenBank, the NCBI protein database, and the literature database PubMed are all summarized under the heading NCBI databases. Concerning research papers, PubMed is the most important source. Patents, brands and firm references are derived from the applicable molecule database itself.

Third, some compounds are common to many species (e.g., astaxanthin) or well researched (e.g., zidovudine). For these reasons, testing such compounds for available literature could result in high numbers of papers and patents.

Remarks

Although biological databases serve primarily as platforms to exchange data between scientists, they are available instruments that: a) contribute to the objectives of the CBD; b) promote justice, effectiveness and research and development; and c) realize the goals of the 'global multilateral benefit-sharing mechanism'. However, they also suffer from various drawbacks.

Objectives of the Convention on Biological Diversity

Publicly available databases are excellent tools to support the benefit-sharing objective of the CBD. Both providers and users of genetic resources could agree to use databases to share some of the central benefits as laid down by the Nagoya Protocol. Such benefits may include:

• Sharing of research and development results (Annex 2(a)): Researchers could make their results and contact details freely available via public databases (whether through deposition of data or publications). Researchers from source states can always contact users and request additional data.
• Contribution to scientific research (Annex 2(b)): Researchers from source states can use the data from databases to conduct their own research.

Table 5.5 Linking the source species to their source states using the GBIF geographical database. The ratio 'Records with source states/All records' roughly indicates how many more source states might exist. The higher the difference, the more specimens there are without locations and therefore the higher the number of possible source states. A '-' indicates absence of the species within the database. Sources: Biodiversity occurrence data accessed through GBIF Data Portal, data.gbif.org, 16 November 2012

Products	Species	Source states	Records with source states/All records
Nereistoxin	Lumbriconereis heteropoda	China, India	3/3
Halogenated furanones	Delisea pulchra	Antarctica, Australia, Falkland Is. (UK), France, Japan, New Zealand	144/161
Dichloroctylisothiazolon	Eunicea sp	Antigua and Barbuda, Bahamas, Barbados, Belize, Bermuda, Brazil, Cayman Is. (UK), Chile, Colombia, Cuba, Dominica, Dominican Rep., Grenada, Guadeloupe, Haiti, Honduras, Jamaica, Martinique, Mexico, Netherland Ant. (NLD), Panama, Peru, Puerto Rico, Saint Lucia, Senegal, Trinidad and Tobago, Turks and Caicos Is. (UK), United States, Venezuela, Virgin Is. (USA)	701/769
Rhamnolipid-2	Pseudomonas aeruginosa	Australia, Japan, Rep. of Korea, New Zealand, Philippines, Poland, Puerto Rico, Uganda, United Kingdom, United States	31/398
Vent$_R$ DNA Polymerase	Thermococcus litoralis	Italy	3/3
Deep Vent$_R$ DNA Polymerase	Pyrococcus sp	Italy, Japan	5/9
Superoxide dismutase	Thermus thermophilus	Japan, Portugal	9/14
Spirulina Pacifica	Spirulina platensis	Argentina, Chad, Japan, Mexico, Netherlands, Peru, United States, Uruguay	28/28
Astaxanthin	Haematococcus pluvialis	Austria, Germany, Portugal, Sweden, United Kingdom, United States, Uruguay	26/69
Docosahexaenoic acid	Crypthecodinium cohnii	Denmark	1/1
	Microciona prolifera	Singapore, United States	289/293
	Sardina ocellata	-	N/a
	Chlorella hirataii	-	N/a
Pseudopterosins	Pseudopterogorgia elisabethae	Bahamas, Cuba, Jamaica, Netherland Ant., United States, Venezuela	16/18
Spongothymidine, spongouridine	Tethya crypta	Bahamas, Jamaica, United States	12/12
	Eunicella cavolini	France, Italy, Monaco	13/16

Products	Species	Source states	Records with source states/All records
ω-conotoxin	*Conus magus*	Australia, Central African Rep., Fiji, France, India, Indonesia, Japan, Kiribati, Madagascar, Malaysia, Marshall Is., Mauritius, Mexico, Fed. St. Micronesia, New Caledonia (FRA), Oman, Palau, Papua New Guinea, Philippines, Seychelles, Singapore, Solomon Is., South Africa, Sri Lanka, Tanzania, Vanuatu, Vietnam, Yemen	801/949
Ecteinascidin-743	*Ecteinascidia turbinata*	Bahamas, Bermuda, Grenada, Jamaica, Mexico, Puerto Rico, Tunisia, United States	57/83
Cefixime	*Cephalosporium acremonium*	Germany, Sri Lanka, Swaziland, United Kingdom, Zimbabwe	7/497
opAFP-GHc	*Salmo salar*	Argentina, Australia, Canada, China, Denmark, Finland, France, Germany, Greenland, Iceland, Ireland, Italy, Netherlands, Norway, Poland, Russian Fed., Senegal, Spain, Sweden, Ukraine, United Kingdom, United States	37K/40K
	Zoarces americanus	Argentina, Canada, Martinique (FRA), Saint Pierre and Miquelon (FRA), United States	6K/9K
	Oncorhynchus tshawytscha	Argentina, Australia, Canada, Chinese Taipei (Taiwan), Germany, Japan, New Zealand, United States	30K/33K
Green fluorescent protein	*Aequorea victoria*	Canada	11/11
Red fluorescent protein	*Discosoma* sp	Barbados, Belize, Bermuda, Brazil, Chinese Taipei (Taiwan), Cook Is., Cuba, Egypt, Fr. Polynesia (FRA), Jamaica, Fed. St. of Micronesia, Netherlands Ant., Oman, Panama, Puerto Rico, Saint Kitts and Nevis, Seychelles, Tanzania, United States, Virgin Is. (UK)	91/112

Table 5.6 Non-exhaustive list of research papers, patents, brands, firms and cross-references which databases contain for biological molecules. '-' indicates that the database does either not apply to the type of molecule, does not contain an entry on the molecule or it does not contain references to that type of monitoring instrument

Compound	DMNP	RÖMPP	Merck Index	NCBI databases	NAPRALERT	UniProt
Nereistoxin	5 papers	4 papers	9 papers	35 papers	2 papers	-
	-	-	2 patents	144 patents	-	
	-	4 brands	1 brand		-	-
	-	-	1 firm		-	
Halogenated furanones	13 papers	-	-	2 papers (general)	2 papers	-
Dichloroctyliso-thiazolon	-	4 papers	4 papers	16 papers		
		-	-	444 patents	-	-
		-	-	1 brand	-	
		-	1 firm	-		
Rhamnolipid-2	12 papers	8 papers	-	18 papers	1 paper	-
Vent$_R$ DNA Polymerase	-	-	-	2 papers	-	2 papers
	-					30 cross-refs
Deep Vent$_R$ DNA Poly.	-	-	-	2 papers	-	1 paper
						34 cross-refs
Astaxanthin	12 papers	7 papers	13 papers	982 papers	80 papers	-
	-	-	-	69 patents	-	
Docosahexaenoic acid	16 papers	5 papers	5 papers	>9K papers	7 papers	
				45 patents		
Pseudopterosins	17 papers	7 papers	-	24 papers	12 papers	-
	1 brand	-		-	-	
Spongothymidine	9 papers	7 papers	-	78 papers	1 paper	-
	-	-		33 patents	-	
Spongouridine	11 papers	7 papers	-	81 papers	2 papers	-
	-	-		39 patents	-	
Acyclovir	-	4 papers	14 papers	>13K papers	-	-
		-	2 patents	>4K patents		
		1 brand	9 brands	-		
		1 firm	9 firms	-		

Compound	DMNP	RÖMPP	Merck Index	NCBI databases	NAPRALERT	UniProt
Zidovudine	-	4 papers	10 papers	>15K papers	-	-
		-	1 patent	>1K patents		
		1 brand	1 brand	-		
		1 firm	1 firm	-		
Cytarabine	-	10 papers	6 papers	>15K papers	-	-
		-	-	>3K patents		
		1 brand	5 brands			
		-	5 firms			
Vidarabine	26 papers	3 papers	12 papers	>4K papers	3 papers	-
	-	-	2 patents	>1K patents	-	
	6 brands	-	2 brands	-	-	-
	-	-	2 firms	-	-	-
ω-conotoxin M VIIA	12 papers	8 papers	7 papers	194 papers	1 paper	12 papers
	1 brand	1 brand	1 brand	-	-	1 brand
	-	-	1 firm	-	-	1 firm
	-	-	-	-	-	13 cross-refs
Ecteinascidine-743	24 papers	13 papers	11 papers	241 papers	5 papers	-
	-	-	2 patents	8 patents	1 patent	
	1 brand	-	1 brand	-	-	
	-	-	1 firm	-	-	
Cefixime	-	9 papers	10 papers	571 papers	31 papers	-
			2 patents			
			6 brands			
			6 firms			
opAFP-GHc	-	-	-	2 papers	-	-
Green fluorescent protein	-	-	-	>39K papers	-	-
Red fluorescent protein	-	-	-	>1K papers	-	-

- Contribution to education and training (Annex 2(d)): Many databases provide educational and supportive documentary material on using their data.
- Access to scientific information relevant to conservation and sustainable use of biological diversity including biological inventories and taxonomic studies (Annex 2(k)): Many databases on compounds that mention the source species (plus GBIF) offer taxonomic services that help to assess threatened species and to manage biodiversity in source states.

Justice

Biological databases are sound scientific means which allow products derived from the utilization of genetic resources to be traced back to (all) the resource's countries of origin. According to Article 2.4 of the CBD, countries of origin are countries 'supplying genetic resources collected from *in-situ* conditions'. Databases can thus complement existing regional agreements, which aim to provide an equitable distribution of benefits between countries of origin but which lack adequate means to unequivocally identify these states. Examples are Decision 391 of the Andean Community, the draft ASEAN Framework Agreement of Access to Biological and Genetic Resources or the Framework Agreement of the Hindu Kush-Himalayan Region.

Applying the databases analysed in this chapter has shown that almost no source species is endemic to one country. Of the 22 species mentioned in Table 5.5, 19 occur in more than one state, the majority have ranges covering two to eight states, and some have ranges covering as many as thirty states. Given the poor knowledge-base which exists on the geographical distribution of species, particularly for small species, it is highly probable that these species occur in even more states than those already identified. Moreover, 13 of these species are not confined to a clear region but are scattered across many regions. It is possible to assume that these species do not exist in isolation from each other and that more specimens of same species exist between these regions, thus covering more source states than previously identified. It is fair to say that continued marine biodiversity research will reveal larger ranges of these species. If states are receptive to transboundary cooperation as envisaged by Article 11 of the Nagoya Protocol and to improve distributional justice in benefit sharing, biological databases provide the most useful means of finding these states. In addition, some regional ABS regimes, such as the Andean common regime on ABS, mention a regional trust fund that distributes funds among member states. GBIF could support the identification of those member states 'eligible' to receive benefits.

Effectiveness

Databases of biological molecules contribute greatly to effective monitoring by referring to the relevant research papers, patents, brands and companies.

Research papers are the main tool for monitoring the largest part of the research and development phase for a particular compound. Before a compound reaches the market, it goes through a long series of tests that acquire basic information on the compound. The number of researchers and institutes involved in this process is unmanageable as illustrated by Table 5.6, since thousands of entities might be involved in research and development – it would be impossible to ensure effective monitoring of this process by overloading pioneer users (access seekers) with reporting duties, comeback clauses and other obligations. Publication lists within databases greatly improve this situation. If the name of the compound is known, these lists provide a good overview of those researchers who have used the compound and the relevant research papers. These research papers can be seen as the final reports which users are commonly obliged to submit in order to ensure monitoring. Providers can access these papers over literature databases, such as PubMed, to obtain all relevant information on the users, uses and benefits generated.

References to patents, brands and even the firms marketing these brands are a powerful tool for monitoring the marketing phase of products. These tools allow providers to identify users with commercial intentions and request that these users share any commercial benefits. Before entering into ABS negotiations, providers can gather as much information as possible to strengthen their bargaining position. For example, most users with commercial intentions are larger companies that are listed on stock exchanges and which therefore have to disclose annual reports. Providers can access these reports, which often contain detailed information about the company and revenues of particular market products.

Product databases in combination with geographical databases, such as GBIF, establish direct links between the user and the source states. Source states can then easily contact users and request commercial and non-commercial benefit sharing. However, source states should not abuse their position and demand unrealistic benefits that cannot be delivered. For example, many researchers cannot share commercial benefits. Benefit sharing must be on mutually agreed terms, otherwise, it could curtail research and development.

A novel and related approach to monitoring could be applying persistent identifiers – unique identification codes that accompany genetic resources downstream (Garrity *et al.*, 2009). As soon as a user collects and intends to use a genetic resource, the resource would receive an identifier that is stored at a central repository. Any transfer, new use or feeding into databases would be recorded by the repository. For monitoring, provider states could use such identifiers to consult the repository, which then provides the locations and uses of the resource. ABS-relevant documents, such as certificates of compliance and patents, could also be linked to the identifier. To make such a system work, specific identifiers must be used universally by all stakeholders, and users need to inform the repository about transfers and new uses. A similar system is already applied by GenBank that employs 'accession numbers' for DNA sequences.

Research and development

Biological databases promote research and development in three ways. First, provider states do not need to adopt overly stringent ABS legislation in order to assure the legitimate utilization of genetic resources. One of the major concerns that prompts source states to adopt overly stringent ABS legislation is losing control over their genetic resources once they have been exported. Biological databases can improve this situation by allowing source states to monitor and therefore retain control over the uses of their genetic resources by using, for example, persistent identifiers. Source states may therefore choose to adopt simplified access legislation and impose fewer monitoring obligations on users (Kamau *et al.*, 2010, p.259). This in turn might attract scientific research.

Second, biological databases promote the exchange of data between scientists by offering greater advantages compared with traditional journals. They first, have much more space; second, concentrate research results from many different researchers and make them available in an aligned and concise way; third, publish data and results much more quickly; fourth, allow the downloading of data for further use; fifth, provide tools and software for data comparison and manipulation and sixth, often offer these services for nothing, the only condition being proper acknowledgement.

Third, biological databases could increase taxonomic and biogeographic research efforts in source states. The more species a state identifies within its boundaries, the higher their chance of being an 'official' source state for a particular product derived from utilization of a shared marine genetic resource. If states agree to share benefits when they share the same genetic resource, the more species a state has identified, the higher are their chances of qualifying for benefit sharing.

The global multilateral benefit-sharing mechanism

One of the most innovative, yet challenging achievements of the Nagoya Protocol is the 'global multilateral benefit-sharing mechanism' mentioned in Article 10. In cases where utilization generates benefits from genetic resources that occur in transboundary situations or for cases where it is not possible to grant or obtain prior informed consent (PIC), this mechanism would use these benefits to support conservation and sustainable use globally. This formulation is deliberately broad and can cover several particular situations depending on the interpretation of this provision.

The benefit-sharing mechanism can contribute to a 'grand global bargain' on biodiversity conservation by complementing bilateral ABS transactions by multilateral ABS transactions in situations that still need to be defined. Regardless of any future interpretations and agreements, such situations could involve a) species whose specimens occur in situ within the jurisdictions of multiple states (transboundary); b) species whose specimens occur ex situ within jurisdictions of multiple states (transboundary); c) collecting species in areas beyond national jurisdiction (no PIC); d) species that migrate between areas within and beyond national jurisdiction (transboundary); e) specimens that are exported and whose biotechnological potential is accessed in a jurisdiction different from the source or provider country (transboundary); f) collecting specimens in countries without ABS legislation (no PIC); g) specimens that were collected before the CBD entered into force (no PIC); h) the user has not obtained PIC or has received material from intermediaries that have not obtained prior informed consent (no PIC); i) negotiations on PIC have failed or prior informed consent has not been provided for any reason (no PIC) or j) products contain genetic resources from so many sources that it is impractible to obtain prior informed consent for each resource (no PIC) (Greiber *et al.*, 2012, p.129; Tvedt, 2012). Since not all situations mentioned above are politically feasible, and this book focuses on the marine environment, the following analysis focuses on marine genetic resources that were collected in maritime zones and examines how such a benefit-sharing mechanism might roughly operate in such circumstances.

MODE OF OPERATION

From the outset, the global orientation might cause some confusion, especially when genetic resources straddle the jurisdictions of multiple states. While 'global' indicates the participation of all states worldwide, as with the common heritage principle, states that partake in the distribution of same genetic resources would more than likely attempt to claim an exclusive part of the benefits. In such cases, the multilateral benefit-sharing mechanisms would quickly compete with the paradigm on bilateral ABS transactions and run into difficulties, as long as a scientific basis for substantiating the distribution of marine genetic resources and thus identifying legitimate beneficiaries is lacking. This would impair the viability of the mechanism and its broader utility. As a solution, biological databases could supply the important infrastructure of the mechanism to ensure effective operation.

The objective of the mechanism – to use benefits to support the conservation of biological diversity and the sustainable use of its components globally – is already described within Article 10 of the Nagoya Protocol. How such benefit sharing could work out in practice depends on the geographic situations under which marine genetic resources were collected.

The geographic scope of the benefit-sharing mechanism could apply to geographically-specific situations that are inadequately captured by the CBD. Because the mechanism covers genetic resources occurring in transboundary situations or for which it is not possible to obtain prior informed consent, it can apply to three distinct situations for in-situ marine genetic resources (presupposing that a country has adopted ABS legislation): a) marine genetic resources occurring in the high seas and the Area (no PIC); b) genetic resources straddling maritime zones belonging to different states (transboundary situations); or c) marine genetic resources occurring in the maritime zones of coastal states and high seas or the Area (transboundary situations and no PIC).

Depending on those situations, the direction of benefit sharing may vary. With regard to genetic resources in the high seas or the Area, benefit sharing may take one of two directions. One option is to channel the benefits into conservation and sustainable use projects in areas beyond national jurisdiction, such as marine protected areas in the high seas managed by regional seas agreements. The other option could be to expressly declare common heritage status for such resources.

Benefits would thus profit humankind at large. They could be distributed by an independent international governing body (CBD ICNP, 2012c, para. 25), in a way similar to the Multilateral System within Articles 10–13 of the ITPGRFA or the Global Environment Facility (GEF, 2010): Benefits would accrue to a central mechanism and would then be distributed to projects worldwide that support the aims of the CBD. The advantage of such an approach is to allocate benefits in areas within national jurisdiction where biodiversity arguably suffers more from anthropogenic pressures than in areas beyond national jurisdiction. A mixture of both options is also feasible.

In cases where a genetic resource occurs in transboundary situations, benefit sharing in a global sense should take another route. As source states would be unlikely to waive benefits generated from genetic resources occurring within their jurisdiction, the mechanism should use a benefit-sharing formula that takes account of these interests. Equal participation in the share of benefits by all source states would be a possible option. An extra share could go to the provider state, to cover any transaction costs that have accumulated when providing the resource. Again, benefits could accrue to a central mechanism first and then be distributed to source states on a project-basis.

When genetic resources occur in areas both beyond and within national jurisdiction, the mechanism would have to develop a formula that blends the above approaches. While one part of benefits would have to go to source states, the other part could either support conservation projects in areas beyond national jurisdiction, flow into a global mechanism or do both.

As already described, the means of identifying the geographical origin of marine genetic resources would be biological databases. As soon an entity seeks to access genetic resources in areas within national jurisdiction and thus requests consent from the competent national authority of the provider state, the authority would use databases to identify other source states. The provider state or the user could then either share benefits directly with the source states, or they could notify the benefit-sharing mechanism and divert a share of benefits to the mechanism, which would then distribute the benefits among source states. The latter approach has the advantage that the mechanism retains the benefits, if the resource occurs in areas beyond national jurisdiction as well. If an entity seeks access in areas beyond national jurisdiction, it should negotiate benefit sharing directly with the mechanism. In such cases, the mechanism would need to use databases to check whether or not the resource also occurs within national jurisdiction. As soon as an entity has produced products from utilization, it should directly notify the benefit-sharing mechanism, which would again use databases to trace the geographical sources and initiate benefit sharing.

It is also imaginable that another institution responsible for using databases for identifying source states could be the new Clearing-House Mechanism under the Nagoya Protocol (CBD ICNP, 2011, pp.7–9). In the context of its mandate of creating information partnerships to attain the goals of the Nagoya Protocol, it might closely collaborate with the benefit-sharing mechanism, by partnering with biological databases to backtrack utilization.

Benefit sharing was originally perceived to support developing states (and states with economies in transition) to meet their conservation targets, and so benefit sharing under the mechanism should also be confined to such states and not include industrialized states. The Global Environment Facility has developed a simple selection scheme in which countries are eligible to receive funds if they are also eligible to borrow from the World Bank or receive technical assistance through the country programme of the United Nations Development Programme (GEF, 2008, para9(b)). These selection criteria could also apply for identifying states eligible to receive funds from the benefit-sharing mechanism.

Last, it is important to understand how to incentivize users and user states to participate in the system. Users are mainly interested in facilitated access, and so the participation of users in the benefit-sharing mechanism could be compensated by provider states through a lifting of ABS requirements. Examples could be less stringent monitoring and benefit-sharing obligations, because these would already be in the hands of the benefit-sharing mechanism. In addition, if users were proactive and participated voluntarily in the benefit-sharing mechanism, it would build trust between users and providers of genetic resources and would thus facilitate ABS transactions in the future.

In conclusion, the benefit-sharing mechanism in combination with biological databases could be a powerful system to support the global conservation of biological diversity. Instead of accumulating benefits only in the provider state, benefits could be distributed among all source states, which better meets the needs of regional biodiversity conservation. However, the mechanism is still in its infancy and needs further development. Although it is beyond the scope of this work to elaborate a detailed structure of the mechanism, the above considerations might provide a stimulus to international negotiations on the development of the mechanism.

Weak points

Despite the advantages of biological databases for supporting ABS, they also have a number of weak points.

IDENTIFYING SINGLE PROVIDER STATES

Biological databases only allow the identification of states that possess a source species in in-situ conditions, i.e., which are according to Article 2.4 of the CBD countries of origin of genetic resources. The actual provider state as defined under Article 2.5 of the CBD, i.e., the country that has supplied the genetic resource collected from in-situ conditions or from ex-situ sources, remains unidentified. States which uphold bilateral exchange of genetic resources and benefits between the user and the provider state, may be especially reluctant to accept this, since they would forgo the full share of benefits.

In response, one must consider the objectives of the CBD, which are the conservation and sustainable use of biodiversity in a *global* sense (CBD Secretariat, 2010). One of the main instruments to achieve these objectives is benefit sharing. This prompts the question: what means of benefit sharing are better suited to promote conservation and sustainable use in a global sense? Biodiversity often straddles national boundaries and has regional distributions; or to put it more bluntly: 'genetic resources are distributed in patterns that represent evolutionary and not political history' (Afreen and Abraham, 2009, p.23). While populations of one species may proliferate in the provider state, other populations of same species may be highly threatened in other source states. From such cases, it becomes obvious that sharing benefits with the provider state alone does not meet the needs of global biodiversity conservation, and that regional or even global sharing of benefits on equal terms better contributes to conservation and sustainable use. The Nagoya Protocol supports this position, when it proposes a global multilateral benefit sharing mechanism and transboundary cooperation in Articles 10 and 11.

FINDING THE RIGHT DATABASE

No single database provides all the services which would allow products to be traced back to their source and research and development to be monitored. Databases vary greatly in their strengths: for example, the Dictionary of Marine Natural Products always mentions the source species, the Merck Index always refers to commercial brands and relevant companies, and NCBI databases are publicly accessible and store large amounts of data, literature (PubMed) and patent references (PubChem). No database combines all of these strengths so far. Thus, a user who identifies a source species via, e.g., the Dictionary of Marine Natural Products, would have to consult PubMed to obtain a broad overview of the research and development process and then consult the Merck Index to find any commercial users. This is problematic because many databases require a subscription fee which is often prohibitively expensive for stakeholders in developing states. This in turn undermines, albeit indirectly, the objectives of the CBD.

However, databases, especially those within the public domain, such as PubChem are subject to rapid development. While PubChem did not refer to any patents in 2011, the database contains today comprehensive references to vast amounts of patents. It is conceivable that in the near future, such databases may combine all relevant ABS references in a publicly-available manner.

UNDERMINING THE CONVENTION ON BIOLOGICAL DIVERSITY

Product databases indirectly constitute ABS situations. They use benefits generated by the utilization of genetic resources, in this case information on biological compounds, and then redistribute this information among interested parties. Some databases do this on a non-commercial basis, while others request a fee for subscription.

However, a fee for accessing a database effectively damages the chance of source states from participating in benefits, which undermines the third objective of the CBD. Simultaneously, such databases generate profits without compensating source states, which arguably creates an unjust situation. To remedy this situation, the owners of commercial databases might choose to enable facilitated access to their database by source states – or developing states in general – by offering reduced fees or providing free licences.

INCOMPLETE DATABASES

No single database can capture the full variety of biological information that exists in the world. Much of our knowledge about biological molecules is fragmented, held by smaller, isolated databases, while there may be many more papers, patents, brands and companies not captured by databases, and much of our taxonomic and biogeographic data remains undisclosed in museums, aquaria and herbaria. In addition, sampling efforts are usually limited, scattered and not standardized, and inventories focus on easily accessible sampling sites and organisms. As a result, biodiversity data are still scarce, biased and sometimes of poor quality. Nevertheless, databases have improved over time and they now offer promising opportunities to improve the current deficiencies in justice and effectiveness of ABS transactions.

OUTDATED BIOGEOGRAPHICAL RECORDS

A more critical point is that records of where a species has been observed can be outdated, especially when they are decades old. That means that a species can be extinct in a country, although it had been observed there earlier. Such cases question the applicability of biogeographical databases for channelling benefits, because countries could be rewarded despite having failed to protect the species within their waters.

TRACING SINGLE COMPOUNDS

States that have provided a compound cannot rely on biological databases alone to effectively monitor the development process – they must also rely on the pioneer user (access seeker). In order to monitor the development process, providers must know the name of the product for which to query the database. However, because users take samples that contain an assortment of many unknown compounds, these names only become available after the isolation and analysis of individual compounds. This means that before databases can become a viable option for monitoring, provider states must allow a user to export the sample and conduct initial analyses on the compounds. Once compounds have been identified and are ready for integration into a database, the user has to supply the name of the compound and other relevant information (e.g., database, accession number, etc.) to the provider state. In such cases, the provider state then has a starting point from which it can monitor further utilization. As a result, export, the initial utilization, and any subsequent reporting remain a critical phase for effective monitoring.

IDENTIFYING TRANSFER BETWEEN USERS

Many providers require users to report on any transfers of material to third parties. However, databases do not allow the tracking of the flow of material, and so any transfers between users remain private. This is not a major problem for most compounds because their development process is documented in the reference lists of databases. However, for some compounds the development process will be kept secret, because users have commercial intentions and want to keep the process secret until the release of the product to the market. In such cases, databases are insufficient to keep track of the process. Nevertheless, once a product is marketed, databases again allow for the tracking of origin.

TECHNICAL BACKGROUND KNOWLEDGE

The search interfaces of many databases require technical background knowledge in order to conduct successful queries. With regard to biological molecules, this often includes knowledge of both common names and synonyms. Because most databases do not have a built-in synonym finder, knowledge about alternative names must be gained from reading the scientific literature. Finding the right keyword becomes more arduous when databases require rare search terms in combination with correct and full spelling. A system of persistent identifiers might ameliorate this deficiency once employed.

Although simple keywords often prove to be sufficient for successful queries, additional search criteria are sometimes necessary. This could include: technical criteria, such as molecular structure; IUPAC names or scientific names; or physical, chemical and biological properties. In a few cases, simple keywords provide too many hits. The search then needs to be narrowed by using additional qualifiers.

INCENTIVE FOR THE RESEARCH COMMUNITY

Databases of biological compounds primarily serve the need of the research community to exchange data and conduct further research. If providers of genetic resources and other ABS-stakeholders start using databases to monitor utilization, the research community might perceive this as an unsolicited intrusion into their domain. Scientists could take a stand against being monitored through databases and stop providing their data, which would impede research and development.

In response, scientists should regard databases as a new tool to facilitate research and development. The current trend in ABS transactions is that providers and provider states are increasingly distrustful of users, which results in cumbersome ABS legislation, in turn impairing research and development. Databases could be a tool to solve this issue by building trust between providers and users. Providers would not need to rely on overly strict ABS legislation to ensure monitoring but could make a convenient switch to databases. This could result in lighter access conditions and relieve scientists from burdensome monitoring obligations. It would also simplify benefit sharing, because the main benefits which scientists can offer, namely research results, are in any case available over the database. In conclusion, if the research community could accept being monitored over databases, it could result in simplified access to genetic resources in other states, while the process of research and development remains largely unaffected.

While the incentive for the research sector to support databases for ABS purposes is obvious, it is less clear for the commercial sector. Commercial users, such as large biotechnology companies, have little direct interest in simplified access procedures, because they rely mainly on researchers with non-commercial intentions or intermediaries to obtain samples and data. Instead, commercial users may choose to block their data on commercial products from entering databases in order to avoid any benefit-sharing claims. Having said this, it would be difficult for commercial users to achieve this aim in practice, because there are often dedicated staff collecting such information. In addition, states could choose to adopt laws that force commercial users to feed commercially-relevant data in databases.

Conclusions

Although biological databases were not originally developed to support the ABS system, they have great potential to improve it. They make major benefits, such as research results and data, publicly available, which contributes to benefit sharing. They allow the identification of all source states sharing the same genetic resource, which is the basis for just participation of source states in access negotiations and benefit sharing in transboundary cooperation or the global multilateral benefit-sharing mechanism. They provide research papers, patent numbers, brands and names of companies, which support effective monitoring. And they also promote research and development, because provider states need not to rely on burdensome access provisions to ensure effective monitoring.

Although biological databases could improve the ABS system in theory, they also have a number of weaknesses: stakeholders on both the provider and user side might reject databases as ABS tools; databases have many flaws and gaps in data coverage; users need to combine several databases to obtain all information; and some databases are too technical or expensive for easy application.

In order to improve the viability of using biological databases for ABS, the following steps are recommended. First, parties to the Nagoya Protocol should, when considering the need for and the design of a global multilateral benefit-sharing mechanism, discuss and acknowledge the important contribution biological databases can make when implementing such a mechanism. Databases could be the central scientific tools of this mechanism, unequivocally identifying the countries of origin which would then benefit from the commercialization of products derived from their genetic resources.

Second, regional ABS agreements should now start to use databases to allocate benefits from their common fund among (member) countries of origin. The successful application of this method in regional systems would illustrate the viability of using databases for ABS and it would also promote their value for global benefit-sharing mechanisms.

Third, provider states should see themselves less as receivers of benefits and more as contributors to global biodiversity conservation. They should therefore waive requesting full shares of any future benefits – a situation unlikely to occur in any case, given the minute fraction of samples that are actually developed into commercially viable products. Instead, they should support the collection of monetary benefits by global mechanisms, which use biological databases to identify countries of origin. This approach could also benefit the provider state in the long run, because other states could follow its example, also agreeing that benefits should flow into the mechanism. This would increase the overall funds available to the mechanism and, in turn, increase the funds that eventually flow back to the provider state (if it qualifies as country of origin). In summary, a global mechanism might ensure a more reliable flow of benefits than waiting in the vain hope of being the provider state for the source of a blockbuster product.

Fourth, a large responsibility also falls to the database providers themselves. Larger databases should integrate smaller, isolated databases into their data network, thereby extending their data coverage. They should offer more services that facilitate the backtracing and monitoring of utilization by including, for example, references to commercial products derived from utilization and their manufacturers. Databases should become more user-friendly and, where relevant, cheaper, in order to be accessible for more ABS stakeholders. It is imperative that databases at least mention the source species and, if possible, the country that provided the material. Similarly, they should oblige users to mention the biological and geographical source of the material and to comply with prior informed consent and mutually agreed terms. To enable this, large institutions that maintain a number of databases, such as NCBI, should take a proactive role and guide database operators and users by developing codes of conduct. Such codes should state how users of genetic resources could enter into contractual obligations with providers to use databases for effective ABS transactions.

Fifth, users both with and without commercial intentions should, when feeding data into databases, always refer to the source species of the material and voluntarily provide data on the provider state. It is also important that users try to agree with providers of genetic resources about using biological databases to ensure monitoring and compliance. This could save users from onerous compliance obligations.

Finally, states should strengthen and promote their taxonomic and biogeographic research bases and feed data into GBIF. States should also acknowledge the important contribution databases could make to ABS and promote further development of these databases by financial support.

The actual utility of biological databases will only become visible through successful application. It is therefore suggested that ABS stakeholders start to apply databases without delay.

6   Conclusions

This work has analysed the: a) importance of marine genetic resources for product development; b) legal status of marine genetic resources in the various maritime zones; c) key concepts related to 'genetic resources'; d) international regimes regulating the use of marine genetic resources; e) implementation of these regimes by typical user states; f) weak points of the ABS regime; and g) potential remedies for these weak points.

Marine genetic resources continue to be an important source of raw material for the development of products. Dozens of commercial products already exist in the market. Moreover, the possibility of discovering novel compounds with potential is unlikely to reduce in the near future, given the plethora of unidentified marine species. However, it would be wrong to see the oceans simply as a source of easily harvestable 'blue gold'. Accessing the marine environment is difficult and typically associated with high costs. Developing products then involves substantial risks and may take decades. For example, most products marketed today were collected years before the CBD entered into force. Despite these drawbacks, it is likely that sampling activities in the marine environment will only intensify.

The legal status of marine genetic resources differs according to the maritime zone. In internal waters, the territorial sea, the EEZ and the continental shelf, they are subject to the sovereignty or sovereign rights of the coastal state. As such, the coastal state has the right to control any kind of exploration and exploitation of marine genetic resources, provided that there are no more specific rules qualifying these rights. In the high seas and the Area, marine genetic resources are res communis. That means that such resources are freely available to all, and no entity, such as other states or international organizations, can control the collection and utilization by another. Because this status could result in disadvantages for biodiversity conservation and benefit sharing, the international community is currently discussing options for changing the legal status. One solution, the regulation of marine genetic resources of the Area within a CBD protocol, was ended with the adoption of the Nagoya Protocol, which has nothing to say on this issue. Current international deliberations focus now on the development of an implementation agreement.

The most important key concept relating to genetic resources is 'utilization of genetic resources' because, being the third objective of the CBD, it triggers benefit sharing. The concept of 'utilization' has remained obscure for many years, because it excluded the utilization of many compounds that would intuitively be thought to trigger benefit sharing, but which were not covered by the definition of 'genetic resources'. This has led to many uncertainties about which activities do trigger benefit sharing. This issue was solved only with adoption of the Nagoya Protocol, by integrating the utilization of compounds not qualifying as genetic resources into the ABS regime. Despite this, the Protocol fails to provide much more guidance on specific criteria or uses that constitute 'utilization', referring only to biotechnology and research and development. In the absence of additional guidance from the Protocol, catalogues that interpret 'utilization' and 'biotechnology' can be consulted. Such catalogues provide specific uses that aggregate utilization into activities which aim to reproduce biological systems, analyse biological molecules and manage biological information.

The key international instruments that regulate the use of marine genetic resources are the ABS regime under the CBD, the Nagoya Protocol and several regimes under UNCLOS. These instruments share many similarities and thus support each other. However, there are also a number of discrepancies. Many discrepancies actually complement gaps in the other conventions. Others may pose challenges to member states, e.g., when strong rights in one convention are limited by restrictions in the other. When it comes to consistent implementation, member states must anticipate these discrepancies and adhere to rules of precedence when establishing ABS rules for marine genetic resources.

The national implementation of both conventions in typical user states is still very fragmented. Comprehensive user measures are often missing, as are provider measures. Nevertheless, the first tentative approaches to regulating user behaviour can be seen. These often involve national patent laws that prescribe a declaration of the origin. Despite this progress, more extensive user measures are essential for a functioning ABS system. User states should adopt stand-alone obligations for users to comply with ABS terms and to share any benefits they derive from utilization. User states should also allow provider states to enforce ABS terms through legal actions.

One of the major weak points of the ABS system is that transactions of genetic resources are exclusively bilateral: one provider state grants access to a 'pioneer' user and in return receives a share of the benefits. Such arrangements are unjust because all other source states sharing the same genetic resources are forced to forgo the benefits derived from utilization, which could lead to competition and an inequitable use of genetic resources. Several international and regional instruments have tried to improve justice by establishing global or regional funds. However, as long as the true geographical distribution of genetic resources remains unidentified, so too are the true source states. Such bilateral arrangements are also ineffective, because the provider state relies exclusively on the pioneer user to ensure any utilization is monitored. This is impossible, because the resource changes hands frequently and often results in products with little similarity to the original resource once accessed. Finally, ineffectiveness, the slowness of user states in adopting user measures, as well as the fear of biopiracy, all propel provider states to adopt overly stringent ABS measures to prevent the illegitimate exploitation of their natural resources.

Potential instruments to remedy these weaknesses are biological databases on biological molecules and geographical distribution of their biological sources. First, biological databases contribute to justice by linking a product derived from the source species to all source states. They could thus constitute the central scientific means for implementing transboundary cooperation and the global multilateral benefit-sharing mechanism, as envisaged by the Nagoya Protocol. By identifying all source states of any particular product, states engaged in transboundary cooperation or the benefit-sharing mechanism could allocate benefits to the full range of source states. Second, biological databases improve effectiveness by referring to the scientific publications, patents, brands and companies relevant to utilization of genetic resources. Thus, both provider and user states can quickly gain an overview on the relevant uses since access of a genetic resource has occurred. If a product is commercialized, then the provider state could approach the relevant user and request benefit sharing. Third, biological databases also promote research and development, because provider states need not rely on overly restrictive access regulations to ensure effective monitoring. Fourth, databases make the core benefits envisaged under the Nagoya Protocol available to the international community and thus contribute to the third objective of the CBD. In conclusion, it is necessary for all stakeholders to acknowledge the important contribution which biological databases could make to a functioning ABS system and to start developing such databases into sound ABS instruments.

Notes

1 Introduction

1 Article 2 of the 1992 Convention on Biological Diversity defines 'genetic resources' as 'any material of plant, animal, microbial or other origin containing functional units of heredity', which is 'of actual or potential value'.

2 Factual background

1 However, May (1992) reached a different extrapolation that estimates deep sea species richness at 500,000.
2 The actual producers of compounds are often microbial symbionts that are intrinsically associated with the host organism.
3 Please note that the first phases in research and development are often of a non-commercial nature. This fact is captured by the concept of 'change of intent', which has complicated ABS negotiations over recent years (CBD COP, 2010a, pp.2–4).
4 Ara-A was later extracted from the gorgonian species, *Eunicella cavolini*.
5 In contrast, geographical limits of terrestrial species may fall together with political boundaries (Dikshit, 2000, p.70; Croteau, 2010, p.12).

3 Access and benefit sharing in the marine realm

1 The Convention on the Territorial Sea and the Contiguous Zone, the Convention on the High Seas, the Convention on the Continental Shelf and the Convention on Fishing and Conservation of the Living Resources of the High Seas.
2 Sovereignty is a description of legal personality comprising supremacy of state power over all activities of all entities within the limits of territorial borders and the independence of this power. Sovereignty over natural resources is the same but only applies to activities relating to natural resources (Elian, 1979, pp.5–13). Similarly, the term 'sovereign rights' are those rights derived from sovereignty of state (Brownlie, 2003, p.107).
3 In paragraph 87 of *Texaco Overseas Petroleum Co. and California Asiatic Oil Co. v the Government of the Libyan Arab Republic* (1977), the arbitrator accepted the legality to nationalize foreign assets of natural resources under the permanent sovereignty principle. In paragraph 143 of *The Government of the State of Kuwait v The American Independent Oil Co* (1982), the tribunal accepted nationalization of a foreign company's concessions under the obligation of appropriate compensation.
4 For example, Article 2.3 of the Convention on Wetlands of International Importance especially as Waterfowl Habitat, preamble 6 of the Convention on the Control of

Transboundary Movements of Hazardous Wastes and Their Disposal, Article 3 and 15.1 CBD, Article 193 UNCLOS and preamble (d) of the International Tropical Timer Agreement.

5 This list refers to the CBD; the United Nations Framework Convention on Climate Change, the United Nations Convention to Combat Desertification in Countries Experiencing Serious Drought and/or Desertification, Particularly in Africa; Part XII UNCLOS; the Stockholm Convention on Persistent Organic Pollutants and the Convention for the Protection of the Ozone Layer.

6 The term 'common concern' was first used by the General Assembly to avoid attribution of 'common heritage' to the global climate (UNGA, 1988, preamble 1; Boyle, 1996, p.40)

7 It should be noted here that the discussion on the physical composition of 'functional units of heredity' distracts attention from the complex nature of genetic resources, which also have an information content, i.e., only the sequence, which is materialized within functional units of heredity (Stoll, 2004, p.77). The importance of the genetic information for the production of compounds has led authors to conclude that genetic resources rather constitute an informational resource and that the definition of genetic resources should be extended by the concept of intangible scientific information about genetic functions (Parry, 2000, p.383; Tvedt and Young, 2007, p.65). Despite the overall importance of genetic information for the ABS system, such a reading from the definition of genetic resources does not seem to be possible. Rather 'units of heredity' means the physical material, i.e., the nucleotides that make up a 'functional unit' on nucleic acids (Winter, 2009, p.22).

8 An example, which illustrates the relationship between basic and applied research, is research on corals. Identification of coral species and ecological interaction of species is basic research. The analysis of physiologically active metabolites of specific coral species and their effect on other species qualifies as applied research, because the results can serve as the basis to investigate coral metabolites for potential medical products.

9 Bioprospecting can be described as the 'exploration of biodiversity for commercially valuable genetic and biochemical resources. It can be defined as the process of gathering information from the biosphere on the molecular composition of genetic resources for the development of new commercial products' (CBD COP, 2000a, para. 6).

10 The term 'stakeholder' is not defined within the Guidelines. However, 'stakeholder' is used very broadly and includes 'Parties, Governments' (introduction); 'users and providers' (11(d) and 16) and 'indigenous and local communities' (16(a)(vi), 26(d) and 43(a)). Their involvement is essential to ensure development and implementation of ABS agreements (17), and stakeholders need to be consulted, their views taken into consideration and their involvement be promoted by providing information and capacity-building (18–20, 27(e), 30).

11 'Fair' can be defined as 'just or appropriate in the circumstances' and 'treating people equally' and 'equitable' as 'fair and impartial' (Compact Oxford English Dictionary, 2008). 'Fair and equitable' varies on a case-to-case basis but can be assessed using different criteria: a) a deal on mutually agreed terms; b) alleviated information imbalances through prior informed consent and openness of transactions; c) stakeholder representation and influence; d) understanding of all parties and sufficient time to make informed decisions; e) periodical renegotiation of agreements; f) knowledge of relative contributions; g) efforts to strengthen the position and rights of weaker parties; h) mutual recognition and affirmation of the other party's perception of 'fair and equitable', taking into consideration differing world-views and i) proportionate distribution of benefits to all stakeholders depending on their degree of contribution (CBD COP, 1996b, p.7; Henne, 1997, p.83; CBD COP, 1998c, paras 20–23; Byström *et al.*, 1999, pp.20–25; Greiber *et al.*, 2012, p85).

12 The range of royalties varies on a case-to-case basis: directly commercializing natural products extracted from organisms may yield royalties between 3–5 per cent; natural products converted into chemical derivatives: 2–3 per cent and complete synthesis of products based on natural products: 0.5–1 per cent (Rosenthal, 1997, p.4; ten Kate and Laird, 2000b, p.67).

5   Biological databases for improving the ABS system

1 NCBI hosts a diverse suite of 40 freely available, interlinked databases that can be grouped into eight broad categories: a) literature databases; b) taxonomy databases; c) DNA and RNA sequence databases; d) protein sequence databases; e) genomic sequence databases; f) databases that correlate genetics and medicine; g) databases on structures and specific domains of proteins and h) chemicals and bioassay databases. Some databases are universal. Others are highly specific by focusing on particular species, biomolecular tools, biological pathways and diseases or services.

2 Much information that has been used for introducing OBIS was retrieved from the older version of the OBIS website (OBIS, 2011). This site has been removed so that this information cannot be directly accessed anymore.

3 The Census of Marine Life was a global, 10-year international collaboration of scientists and other stakeholders to assess and explain diversity, distribution and abundance of marine organisms throughout the world's oceans (O'Dor *et al.*, 2010; Williams *et al.*, 2010).

4 Academy of Natural Sciences OBIS Mollusc Database, NMNH Invertebrate Zoology Collections, Australian Institute of Marine Science, Ocean Genome Resource.

5 Sharing of research results (Annex, paragraph 2(a) Nagoya Protocol), support and training in using the database (paragraph 2(d)), availability of educational material (paragraph 2(d)), admittance to databases (paragraph 2(e)) and access to information relevant to conservation and sustainable use including biological inventories (paragraph 2(k)).

6 Ocean Biogeographic Information System, Australian Museum, Biologiezentrum Linz, Bernice Pauahi Bishop Museum, California Academy of Sciences, Museo Argentino de Ciencias Naturales, Senckenberg, Field Museum of Natural History, National Museum of Nature and Science in Japan, GBIF-Sweden, Academy of Natural Sciences, Florida Museum of Natural History, Museum national d'Histoire naturelle, GBIF Spain, Museum of Comparative Zoology at Harvard University, Netherlands Biodiversity Information Facility, National Museum of Natural History at Smithsonian Institution, Marine Science Institute, Yale University Peabody Museum, Queen Victoria Museum and Art Gallery, Royal Belgian Institute of Natural Sciences, Sam Noble Oklahoma Museum of Natural Sciences, SynTax, Arctos.

References

Abiad, N. (2008) *Sharia, Muslim states and international human rights treaty obligations: a comparative study*, British Institute of International and Comparative Law, London

Adams, G. (2010) 'Fast growing salmon cleared as fit for human consumption', www.independent.co.uk/life-style/food-and-drink/news/fast-growing-salmon-cleared-as-fit-for-human-consumption-in-us-2071295.html, accessed 6 November 2012

Adede, A.O. (1979) 'United Nations efforts toward the development of an environmental code of conduct for states concerning harmonious utilization of shared natural resources', *Albany Law Review*, vol. 43, pp.488–512

Afreen, S. and Abraham, B.P. (2008) 'Bioprospecting: promoting and regulating access to genetic resources and benefit sharing', Indian Institute of Management Calcutta Working Paper No. 631, http://facultylive.iimcal.ac.in/sites/facultylive.iimcal.ac.in/files/WPS-631_1.pdf, accessed 6 November 2012

Ali, S. and Llewellyn, C. (2009) 'Marine chemical and medicine resources', in J.H. Steele, K.K. Turekian and S.A. Thorpe (eds) *Encyclopedia of ocean sciences*, Elsevier, Amsterdam

Allem, A.C. (2000) 'The terms genetic resource, biological resource, and biodiversity examined', *The Environmentalist*, vol. 20, pp.335–341

Amaral-Zettler, L., Artigas, L.F., Baross, J., Bharati, P.A.L., Boetius, A. Chandramohan, D., Herndl, G., Kogure, K., Neal, P., Pedrós-Alió, C., Ramette, A., Schouten, S., Stal, L., Thessen, A., de Leeuw, J. and Sogin, M.L. (2010) 'A global census of marine microbes', in A.D. McIntyre (ed.) *Life in the world's oceans. Diversity, distribution, and abundance*, Wiley-Blackwell, Chichester, UK

Apiletti, D., Bruno, G., Ficarra, E. and Baralis, E. (2006) 'Data cleaning and semantic improvement in biological databases', *Journal of Integrative Bioinformatics*, vol. 3, no. 2, pp.1–11

Apodaca, R. (2012) 'Simple CAS number lookup with PubChem', depth-first.com/articles/2007/05/21/simple-cas-number-lookup-with-pubchem, accessed 15 November 2012

Apweiler, R., Marine, M.J., O'Donovan, C. and Pruess, M. (2003) 'Managing core resources for genomics and proteonomics', *Pharmacogenomics*, vol. 4, no. 3, pp.343–350

AquaBounty Technologies (2012) 'AquAdvantage® fish', www.aquabounty.com/products/products-295.aspx, accessed November 2012

Arico, S. (2006) 'The last frontier', *A World of Science*, vol. 4, no. 2, pp.19–23

Arico, S. and Salpin, C. (2005) 'Bioprospecting of genetic resources in the deep seabed', Industrial Biotechnology, vol. 1, no. 4, pp.260–282

Arnaud-Haond, S., Arrieta, J.M. and Duarte, C.M. (2011) 'Marine biodiversity and gene patents', *Science*, vol. 331, pp.1521–1522

Attard, D.J. (1987) *The exclusive economic zone in international law*, Oxford University Press, Oxford

Attorney-General's Department (1980) *Offshore constitutional settlement. A milestone in co-operative federalism*, Australian Government Publishing Service, Canberra

Australian Government (2009) *Australia's fourth national report to the United Nations Convention on Biological Diversity*, Australian Government, Canberra

Australian Government (2012) 'Biological resources access agreement between the Commonwealth of Australia and The Craig J. Venter Institute', www.environment.gov.au/biodiversity/publications/access/braa.html, accessed 12 November 2012

Bains, W. (2004) *Biotechnology from A to Z*, Oxford University Press, Oxford

Banat, I.M., Makker, R.S. and Cameotra, S.S. (2000) 'Potential commercial applications of microbial surfactants', *Applied Microbiology and Biotechnology*, vol. 53, pp.495–508

Baslar, K. (1998) *The concept of the common heritage of mankind in international law*, Kluwer Law International, The Hague

Bayat, A. (2002) 'Science, medicine, and the future. Bioinformatics', *British Medicine Journal*, vol. 324, pp.1018–1022

Beattie, A.J., Barthlott, W., Elisabetsky, E., Farrel, R., Kheng, C.T. and Prance, I. (2005) 'New products and industries from biodiversity', in R. Hassan, R. Scholes and N. Ash (eds) *Ecosystems and human well-being: current state and trends*, vol. 1, Island Press, Washington DC

Benoît, G. and College, S. (2006) 'Bioinformatics', *Annual Review of Information Science and Technology*, vol. 1, pp.179–218

Benson, D.A., Karsch-Mizrachi, I., Lipman, D.J., Ostell, J. and Sayers, E.W. (2010) 'GenBank', *Nucleic Acids Research*, vol. 38, pp.D46–D51

Bergmann, W. and Burke, D.C. (1955) 'Contributions to the study of marine products. XXXIX. The nucleosides of sponges. III. Spongothymidine and spongouridine', *Journal of Organic Chemistry*, vol. 20, pp.1501–1507

Berne Declaration (2010) *Dirty business for clean skin. – Nestlé's rooibos robbery in South Africa*, Berne Declaration Briefing Paper, Zürich

Bhadury, P. and Wright, P.C. (2004) 'Exploitation of marine algae: biogenic compounds for potential antifouling applications', *Planta*, vol. 219, pp.561–578

Bhakuni, D.S. and Rawat, D.S. (2005) *Bioactive marine natural products*, Anamaya Publishers, New Delhi

Biber-Klemm, S. (2008) 'Access to genetic resources and the fair and equitable sharing of the benefits resulting from their use – the challenges of a new concept', *Environmental Law Network International*, vol. 1, pp.2–18

Biber-Klemm, S. and Martinez, S.I. (2009) *Access and benefit sharing. Good practice for academic research on genetic resources*, Swiss Academy of Sciences, Bern

Biber-Klemm, S., Martinez, S.I. and Jacob, A. (2010) *Access to genetic resources and sharing of benefits. ABS programm 2003 to 2010*, Swiss Academy of Sciences, Bern

Bidartondo, M.I. (2008) 'Preserving accuracy in GenBank', *Science*, vol. 319, p.1616

Birnie, P. and Boyle, A.E. (2002) *International law and the environment*, Oxford University Press, Oxford

Birnie, P., Boyle, A.E. and Redgwell, B. (2009) *International law and the environment*, Oxford University Press, Oxford

Blaustein, R.J. (2010) 'High-seas biodiversity and genetic resources: science and policy questions', *BioScience*, vol. 60, no. 6, pp.408–413

Blunt, J.W. and Munro, M.H.G. (2008) *Dictionary of Marine Natural Products with CD-ROM*, Chapman & Hall/CRC, Boca Raton, FL

Blunt, J.W., Copp, B.R., Munro, M.H.G., Northcote, P.T. and Prinsep, M.R. (2010) 'Marine natural products', *Natural Products Report*, vol. 27, pp.165–237

Boetius, A. and Boetius, H. (2011) *Das dunkle Paradies*, Bertelsmann, Munich

Bollmann, M., Bosch, T., Coljin, F., Ebinghaus, R., Froese, R., Güssow, K., Khalilian, S., Krastel, S., Körtzinger, A., Langenbuch, M., Latif, M., Matthiessen, B., Melzner, F., Oschlies, A., Petersen, S., Proelß, A., Quaas, M., Reichenbach, J., Requate, T., Reusch, T., Rosenstiel, P., Schmidtz, J.O., Schrottke, K., Sichelschmidt, H., Siebert, U., Soltwedel, R., Sommer, U., Stattegger, K., Sterr, H., Sturm, R., Treude, T., Vafeidis, A., van Bernem, C., van Beusekom, J., Voss, R., Visbeck, M., Wahl, M., Wallmann, K. and Weinberger, F. (2010) *World ocean review*, maribus, Hamburg, Germany

Bolton, E.E., Wang, Y., Thiessen, P.A. and Bryant, S.H. (2008) 'PubChem: integrated platform of small molecules and biological activities', *Annual Reports in Computational Chemistry*, pp.217–241

Borém, A., Santos, F.R. and Bowen, D.E. (2003) *Understanding Biotechnology*, Prentice Hall, Upper Saddle River, NJ

Boyle, A.E. (1996) 'The Rio Convention on Biological Diversity', in C. Redgwell and M. Bowman (eds) *International law and the conservation of biological diversity*, Kluwer Law International, London

Boyle, A.E. (1997) 'Dispute settlement and the Law of the Sea Convention: problems of fragmentation and jurisdiction', *International and Comparative Law Quarterly*, vol. 46, pp.37–54

Boyle, A.E. (1999) 'Some reflections on the relationship of treaties and soft law', *International and Comparative Law Quarterly*, vol. 48, pp.901–913

Bragg, J.R., Prince, R.C., Harner, E.J. and Atlas, R.M. (1994) 'Effectiveness of bioremediation for the *Exxon Valdez* oil spill', *Nature*, vol. 31, pp.413–418

Brand, U. and Görg, C. (2001) *Access and benefit sharing. Zugang und Vorteilsausgleich – das Zentrum des Konfliktsfelds Biodiversität*, Forum Umwelt und Entwicklung/Germanwatch, Bonn

Briggs, J.C. (1994) 'Species diversity: land and sea compared', *Systematic Biology*, vol. 43, no. 1, pp.130–135

Brown, E.D. (1984) *Seabed energy and mineral resources and the law of the sea. Vol. 1: The areas within national jurisdiction*, Graham & Trotham, London

Brown, E.D. (1994) *The international law of the sea, Volume I, Introductory Manual*, Dartmouth, Aldershot, UK

Brownlie, I. (2003) *Principles of public international law*, Oxford University Press, Oxford

Brundtland, G.H. (1987) *Our common future*, Oxford University Press, Oxford

Bucher, S. (2008) *Der Schutz von genetischen Ressourcen und indigenem Wissen in Lateinamerika*, Nomos, Baden-Baden, Germany

Buck, E.H. (2007) 'U.N. Convention on the Law of the Sea: living resources provisions', in M.B. Paulsen (ed.) *Law of the Sea*, Nova Science Publishers, New York

Bugni, T.S., Richards, B., Bhoite, L., Cimbora, D., Harper, M.K. and Ireland, C.M. (2008) 'Marine natural products libraries for high-throughput screening and rapid drug discovery', *Journal of Natural Products*, vol. 71, pp.1095–1098

Burton, G. (2004) 'Derivatives', Discussion paper presented at the International Expert Workshop on Access to Genetic Resources and Benefit Sharing, Cuernavaca, Mexico

Burton, G. (2009) 'Australian ABS law and administration – a model law and approach?', in E.C. Kamau and G. Winter (eds) *Genetic resources, traditional law and the law. Solutions for access and benefit sharing*, Earthscan, London

Butler, M.S. (2004) 'The role of natural product chemistry in drug discovery', *Journal of Natural Products*, vol. 67, pp.2141–2153

Byström, M., Einarsson, P. and Nycander, G.A. (1999) 'Fair and equitable. Sharing the benefits from use of genetic resources and traditional knowledge', Swedish Scientific Council on Biological Diversity report

Californian Council of Chief Librarians (2004) 'The Merck Index', www.cclibraries.org/reviews/Documents/ear_rev_merckindex.pdf, accessed 14 November 2012

CambridgeSoft (2012) 'Scientific databases. The Merck Index', www.cambridgesoft.com/databases/details/?fid=45, accessed 14 November 2012

CAMERA (Community Cyberinfrastructure for Advanced Marine Microbial Ecology Research and Analysis) (2012) 'Launching the Global Community Cyberinfrastructure for Advanced Marine Microbial Research and Analysis (CAMERA)', http://camera.calit2.net/news/article_cyberinfrastructure.shtm, accessed 12 November 2012

Carr, M.H., Neigel, J.E., Estes, J.A., Andelman, S., Warner, R.R. and Largier, J.L. (2003) 'Comparing marine and terrestrial ecosystems: implications for the design of coastal marine reserves', *Ecological Applications*, vol. 13, no. 1, pp.S90–S107

Carté, B.K. (1996) 'Biomedical potential of marine natural products', *BioScience*, vol. 46, no. 4, pp.271–286

Cataldi, G. (2006) 'Biotechnology and marine biogenetic resources: the interplay between UNCLOS and the CBD', in F. Francioni and T. Scovazzi (eds) *Biotechnology and international law*, Hart Publishing, Oxford

CBD (Convention on Biological Diversity) (2006) 'Elements for a multi-year plan of action for South-South cooperation on biodiversity for development', CBD document UNEP/CBD/BM-SCC/1/2/Rev.2

CBD (2012) 'Database on ABS measures', www.cbd.int/abs/measures, accessed 12 November 2012

CBD ABS GTLE (Group of Technical and Legal Experts) (2008a) 'Concepts, terms, working definitions and sectoral approaches relating to the international regime on access and benefit sharing', CBD ABS GTLE information document UNEP/CBD/ABS/GTLE/1/INF/2

CBD ABS GTLE (2008b) 'Good business practices and case-studies on biodiversity', CBD ABS GTLE information document UNEP/CBD/ABS/GTLE/1/INF/1

CBD COP (Conference of the Parties) (1995a) 'Report of the second meeting of the conference of the parties to the Convention on Biological Diversity', CBD COP official document UNEP/CBD/COP/2/19

CBD COP (1995b) 'Conservation and sustainable use of marine and coastal biological diversity', CBD COP decision UNEP/CBD/COP/II/10

CBD COP (1995c) 'Intellectual Property Rights and Transfer of Technologies Which Make Use of Genetic Resources', CBD COP official document UNEP/CBD/COP/2/17

CBD COP (1995d) 'Access to genetic resources and benefit-sharing: legislation, administrative and policy information', CBD COP official document UNEP/CBD/COP/2/13

CBD COP (1996a) 'Access to genetic resources', CBD COP official document UNEP/CBD/COP/3/20

CBD COP (1996b) 'Fair and equitable sharing of benefits arising from the use of genetic resources', CBD COP information document UNEP/CBD/COP/3/Inf.53

CBD COP (1998a) 'Conservation and sustainable use of marine and coastal biological diversity, including a programme of work', CBD COP decision UNEP/CBD/COP/IV/5

CBD COP (1998b) 'Access and benefit-sharing', CBD COP decision UNEP/CBD/COP/IV/8

CBD COP (1998c) 'Measures to promote and advance the distribution of benefits from biotechnology in accordance with article 19', CBD COP official document UNEP/CBD/COP/4/21

CBD COP (1998d) 'Addressing the fair and equitable sharing of the benefits arising out of genetic resources: options for assistance to developing country parties to the convention on biological diversity', CBD COP official document UNEP/CBD/COP/4/22

CBD COP (1999) 'Report of the panel of experts on access and benefit-sharing', CBD COP official document UNEP/CBD/COP/5/8

CBD COP (2000a) 'Progress report on the implementation of the programmes of work-information on marine and coastal genetic resources, including bioprospecting', CBD COP information document UNEP/CBD/COP/5/INF/7

CBD COP (2000b) 'Access and benefit-sharing', CBD COP decision UNEP/CBD/COP/V/26

CBD COP (2000c) 'Information on marine and coastal genetic resources, including bioprospecting', CBD COP information document UNEP/CBD/COP/5/INF/7

CBD COP (2002) 'Access and benefit-sharing as related to genetic resources', CBD COP decision UNEP/CBD/COP/VI/24

CBD COP (2004) 'Access and benefit-sharing as related to genetic resources', CBD COP decision UNEP/CBD/COP/DEC/VII/19

CBD COP (2006) 'Marine and coastal biological diversity: conservation and sustainable use of deep seabed genetic resources beyond the limits of national jurisdiction', CBD COP decision UNEP/CBD/COP/VIII/21

CBD COP (2010a) 'Access to genetic resources and the fair and equitable sharing of benefits arising from their utilization', CBD COP decision UNEP/CBD/COP/DEC/X/1

CBD COP (2010b) 'Report of a scientific experts meeting on access and benefit sharing in non-commercial biodiversity research', CBD COP information document UNEP/CBD/COP/10/INF/43

CBD EP-ABS (Panel of Experts on ABS) (1999) 'Options for access and benefit-sharing arrangements', CBD EP-ABS document UNEP/CBD/EP-ABS/2

CBD ICNP (Intergovernmental Committee for the Nagoya Protocol) (2011) 'Report of the expert meeting on the modalities of operation of the access and benefit-sharing clearing-house', CBD ICNP official document UNEP/CBD/ICNP/1/2

CBD ICNP (2012a) 'Measures to raise awareness of the importance of the genetic resources and associated traditional knowledge, and related access and benefit-sharing issues', CBD ICNP recommendation UNEP/CBD/ICNP/2/6

CBD ICNP (2012b) 'Cooperative procedures and institutional mechanisms to promote compliance with the protocol and to address cases of non-compliance', CBD ICNP recommendation UNEP/CBD/ICNP/2/7

CBD ICNP (2012c) 'Synthesis of views with respect to the need for and modalities of a global multilateral benefit-sharing mechanism (Article 10)', CBD ICNP official document UNEP/CBD/ICNP/2/7

CBD SBSTTA (Subsidiary Body of Scientific, Technical and Technological Advice) (1996a) 'Bioprospecting of genetic resources of the deep sea-bed', CBD SBSTTA official document UNEP/CBD/SBSTTA/2/15

CBD SBSTTA (1996b) 'Ways and means to promote and facilitate access to, and transfer and development of technology, including biotechnology', CBD SBSTTA official document UNEP/CBD/SBSTTA/2/6

CBD SBSTTA (2003a) 'Marine and coastal biodiversity: review, further elaboration and refinement of the programme of work. Study of the relationship between the Convention on Biological Diversity and the United Nations Convention on the Law of the Sea with regard to the conservation and sustainable use of genetic resources on the deep seabed', CBD SBSTTA information document UNEP/CBD/SBSTTA/8/INF/3/Rev.1

CBD SBSTTA (2003b) 'Marine and coastal biodiversity: review, further elaboration and refinement of the programme of work', CBD SBSTTA official document UNEP/CBD/SBSTTA/8/9/Add.3/Rev.1

CBD SBSTTA (2003c) 'Study of the relationship between the Convention on Biological Diversity and the United Nations Convention on the Law of the Sea with regard to the conservation and sustainable use of genetic resources on the deep seabed', CBD SBSTTA information document UNEP/CBD/SBSTTA/8/INF/3/Rev. 1

CBD SBSTTA (2012) 'Implementation of the Nagoya Protocol on Access and Benefit-Sharing and the Global Taxonomy Initiative', CBD SBSTTA information document UNEP/CBD/SBSTTA/16/INF/37

CBD Secretariat (2010) *Global Biodiversity Outlook 3*, CBD, Montreal

CBD WG-ABS (Working Group on Access and Benefit-sharing) (2001) 'Regulating access and benefit sharing: basic issues, legal instruments, policy proposals', CBD WG-ABS information document UNEP/CBD/WG-ABS/1/INF/4

CBD WG-ABS (2004) 'Analysis of existing national, regional and international legal instruments relating to access and benefit-sharing and experience gained in their implementation, including identification of gaps', CBD WG-ABS official document UNEP/CBD/WG-ABS/3/2

CBD WG-ABS (2005a) 'Access and benefit-sharing of genetic resources', CBD WG-ABS information document UNEP/CBD/WG-ABS/4/INF/10

CBD WG-ABS (2005b) 'The commercial use of biodiversity: an update on current trends in demand for access to genetic resources and benefit-sharing, and industry perspectives on ABS policy and implementation', CBD WG-ABS information document UNEP/CBD/WG-ABS/4/INF/5

CBD WG-ABS (2005c) 'Analysis of existing national, regional and international legal instruments relating to access and benefit sharing and experience gained in their implementation, including identification of gaps', CBD WG-ABS official document UNEP/CBD/WG-ABS/3/2

CBD WG-ABS (2007) 'Compilation of submissions provided by parties, governments, indigenous and local communities and stakeholders on concrete options on substantive items on the agenda of the fifth and sixth meetings of the ad hoc open-ended working group on access and benefit-sharing', CBD WG-ABS information document UNEP/CBD/WG-ABS/6/INF/3

CBD WG-ABS (2008) 'Report of the meeting of the group of legal and technical experts on concepts, terms, working definitions and sectoral approaches' CBD WG-ABS official document UNEP/CBD/WG-ABS/7/2

CBD WG-ABS (2009) 'Study on the relationship between an international regime on access and benefit sharing and other international instruments and forums that govern the use of genetic resources – the Antarctic Treaty System (ATS) and the United Nations Convention on the Law of the Sea (UNCLOS)' CBD WG-ABS information document UNEP/CBD/WG-ABS/7/INF/3/Part.3

CBD WG-ABS (2010) 'The concept of "genetic resources" in the Convention on Biological Diversity and how it relates to a functional international regime on access and benefit-sharing' CBD WG-ABS information document UNEP/CBD/WG-ABS/9/INF/1

Center for Information Technology (2012) 'NIH Roadmap', http:// cit.nih.gov/ServiceCatalog/NihRoadMap.htm, accessed 15 November 2012

Cermeño, P., de Vargas, C., Abrantes, F. and Falkowski, P.G. (2010) 'Phytoplankton biogeography and community stability in the ocean', *PLoS ONE*, vol. 5, no. 4, pp.1–7

Chalfie, M., Tu, Y., Euskirchen, G., Ward, W.W. and Prasher, D.C. (1994) 'Green fluorescent protein as a marker for gene expression', *Science*, vol. 263, pp.802–805

Chambers, W.B. (2003) 'WSSD and an International Regime on Access and Benefit Sharing: Is a protocol the appropriate legal instrument?', *RECIEL*, vol. 12, no. 3, pp.310–320

Changyun, W., Haiyan, L., Changlun, S., Yanan, W., Liang, L. and Huashi, G. (2008) 'Chemical defensive substances of soft corals and gorgonians', *Acta Ecologica Sinica*, vol. 28, no. 5, pp.2320–2330

Charles, T. (2005) 'Final report', GloFish LLC Report 2005

CHEMnetBase (2012a) 'Introduction to the Dictionary of Natural Products online', http:// dnp.chemnetbase.com/intro/index.jsp, accessed 14 November 2012

CHEMnetBASE (2012b) 'Welcome to CHEMnetBASE', www.chemnetbase.com, accessed 14 November 2012

CHEMnetBASE (2012c) 'Dictionary of marine natural products', www.dmnp. chemnetbase.com/intro/index.jsp#introduction, accessed 6 November 2012

Chin, D., Boyle, G.M., Theile, D.R., Parsons, P.G. and Coman, W.B. (2006) 'The human genome and gene expression profiling', *Journal of Plastic, Reconstructive and Aesthetic Surgery*, vol. 59, pp.902–911

Chiou, P.P., Khoo, J., Chun, C.Z. and Chen, T.T. (2007) 'Transgenic fish', in R.A. Meyers (ed) *Genomics and genetics*, Wiley-VCH, Weinheim

Chishakwe, N. (2009) 'SADC: access to genetic resources, and sharing the benefits of their use – international and sub-regional issues', in T. Young (ed.) *Covering ABS: addressing the need for sectoral, geographical, legal and international integration in the ABS regime*, IUCN, Gland, Switzerland

Chishakwe, N. and Young, T.R. (2003) 'Access to genetic resources and sharing the benefits of their use: international and sub-regional issues', Paper presented at the priority-setting workshop for the Southern Africa Biodiversity Support Programme

Cicin-Sain, B., Knecht, R.W., Bouman, L.D. and Fisk, G.W. (1996) 'Emerging policy issues in the development of marine biotechnology', *Ocean Yearbook*, vol. 12, pp.179–206

Claverie, J.-M. and Notredame, C. (2007) *Bioinformatics for dummies*, Wiley, Hoboken, NJ

Cohen, Y. (2002) 'Bioremediation of oil by marine microbial mats', *International Microbiology*, vol. 5, pp.189–193

Cole, A. (2005) 'Looking for new compounds in sea is endangering ecosystem', *British Medical Journal*, vol. 330, p.1350

Colwell, R.R. (2002) 'Fulfilling the promise of biotechnology', *Biotechnology Advances*, vol. 20, pp.215–228

Commoner, B. (2002) 'Unraveling the DNA myth. The spurious foundation of genetic engineering', *Harper's Magazine*, pp.39–47

Compact Oxford English Dictionary (2008), 3rd ed., Oxford University Press, New York

Concise Oxford English Dictionary (2006), Oxford University Press, New York

Cook, J.T., McNiven, M.A., Richardson, G.F. and Sutterlin, A.M. (2000) 'Growth rate, body composition and feed digestibility / conversion of growth-enhanced transgenic Atlantic salmon (*Salmo salar*)', *Aquaculture*, vol. 188, pp.15–32

Cooper, D. (1993) 'The International Undertaking on Plant Genetic Resources', *RECIEL*, vol. 2, no. 2, pp.1–16

Cooper, H.D. (2002) 'The International Treaty on Plant Genetic Resources for Food and Agriculture', *RECIEL*, vol. 11, no. 1, pp.1–16

Copping, L. (2001) *The biopesticide manual: a world compendium*, British Crop Protection Council, Farnham

Cordell, G.A. (2000) 'Biodiversity and drug discovery – a symbiotic relationship', *Phytochemistry*, vol. 55, pp.436–480

Correa, C.M. (2004) 'Access to genetic resources and the FAO treaty. The case of Andean countries', *Journal of International Biotechnology Law*, vol. 1, pp.153–155

Costello, M.J. (2006) Personal communication

CRC Press (2012) 'Home site', www.crcpress.com/index.jsf, accessed 14 November 2012

CRCnetBASE (2012) 'Welcome to CRCnetBASE', www.crcnetbase.com, accessed 14 November 2012

Croteau, E.K. (2010) 'Causes and consequences of dispersal in plants and animals', *Nature Education Knowledge*, vol. 1, no. 11, p.12

Cyanotech (2012) *Annual Report*, www.cyanotech.com/pdfs/Cyanotech_2012_Annual_Report.pdf, accessed 6 November 2012

Das, P., Mukherjee, S. and Sen, R. (2009) 'Biosurfactant of marine origin exhibiting heavy metal remediation properties', *Bioresource Technology*, vol. 100, pp.4887–4890

Dávalos, L.M., Sears, R.R., Raygorodetsky, G., Simmons, B.L., Cross, H., Grant, T., Barnes, T., Putzel, L. and Perzecanski, A.L. (2003) 'Regulating access to genetic resources under the Convention on Biological Diversity: an analysis of selected case studies', *Biodiversity and Conservation*, vol. 12, pp.1511–1524

Dayan, F.E., Cantrell, C.L. and Duke, S.O. (2009) 'Natural products in crop protection', *Bioorganic & Medicinal Chemistry*, vol. 17, pp.4022–4033

de Fontaubert, A.C., Downes, D.R. and Agardy, T.S. (1996) *Biodiversity in the seas: implementing the Convention on Biological Diversity in marine and coastal habitats*, IUCN, Gland

de Jonge, B. and Louwaars, N. (2009) 'The diversity of principles underlying the concept of benefit sharing' in E.C. Kamau and G. Winter (eds) *Genetic resources, traditional knowledge and the law. Solutions for access and benefit sharing*, Earthscan, London

de la Calle, F. (2009) 'Marine genetic resources. A source of new drugs. The experience of the biotechnology sector', *The International Journal of Marine and Coastal Law*, vol. 24, pp.209–220

de Nys, R. and Steinberg, P.D. (2002) 'Linking marine biology and biotechnology', *Current Opinion in Biotechnology*, vol. 13, pp.244–248

Deckwer, W.-D., Dill, B., Eisenbrand, G., Fugmann, B., Heiker, F.R., Hulpke, H., Kirschning, A., Pohnert, G., Pühler, A., Schmid, R.D. and Schreier, P. (2005) *Autorenhinweise RÖMPP Online*, Thieme Chemistry, Stuttgart

Dedeurwaerdere, T., Iglesia, M., Weiland, S. and Halewood, M. (2009) *The use and exchange of microbial genetic resources for food and agriculture*, FAO, Rome

Deming, J.W. (1998) 'Deep ocean environmental biotechnology', *Current Opinion in Biotechnology*, vol. 9, pp.283–287

Demunshi, Y. and Chugh, A. (2010) 'Role of traditional knowledge in marine bioprospecting' *Biodiversity Conservation*, vol 19, pp.3015–3033

Department of the Environment and Heritage (2005) *Genetic resources management in commonwealth areas*, Australian Government, Canberra

Department of the Environment, Sport and Territories (1996) *National Strategy for the Conservation of Australia's Biological Diversity*, Australian Government, Canberra

Desai, J.D. and Banat, I.M. (1997) 'Microbial production of surfactants and their commercial potential', *Microbiology and Molecular Biology Reviews*, vol. 61, no. 1, pp.47–64

Dewar, A. (2005) *Agrow's top 20: 2005 edition*, T&F Informa UK Ltd, London

DEWHA (Department of the Environment, Water, Heritage and the Arts) (2012a) 'Permits – accessing biological resources in Commonwealth areas', www.environment.gov.au/biodiversity/science/access/permits/index.html, accessed 12 November 2012

DEWHA (2012b) 'Model access and benefit-sharing agreements', www.environment.gov.au/biodiversity/science/access/model-agreements/index.html', accessed 12 November 2012

DFG (German Research Foundation) (2010) *Guidelines for funding proposals concerning research projects within the scope of the Convention on Biological Diversity*, DFG, Bonn

Dictionary of Marine Natural Products (2012) 'Private policy', dmnp.chemnetbase.com/chcd_help/index.jsp?sec=0:28&id=29, accessed 14 November 2012

Dikshit, R.D. (2000) *Political geography. The spatiality of politics*, Tata McGraw-Hill, New Delhi

Division of Program Coordination, Planning and Strategic Initiatives (2012) 'Molecular libraries and imaging', http://nihroadmap.nih.gov/molecularlibraries/index.aspx, accessed 15 November 2012

Dow Jones (2010) 'IMS Health: 2010 global pharma sales to rise 4%–6%', www.dowjones.de/site/2009/10/ims-health-2010-global-pharma-sales-to-rise-46.html, accessed 6 November 2012

Dross, M. and Wolff, F. (2005) *New elements of the International Regime on Access and Benefit sharing of Genetic Resources – the role of certificates of origin*, Bundesamt für Naturschutz, Bonn

Dunham, R.A. (2004) *Aquaculture and fisheries biotechnology*, CABI Publishing, Oxfordshire

Dutfield, G. (2002) 'Sharing the benefits of biodiversity. Is there a role for the patent system?', *The Journal of World Intellectual Property*, vol. 5, no. 6, pp.899–931

Dworjanyn, S.A., de Nys, R. and Steinberg, P.D. (2006) 'Chemically mediated antifouling in the red alga *Delisea pulchra*', *Marine Ecology Progress Series*, vol. 318, pp.153–163

EC (European Community) (1998) 'Communication from the Commission to the Council and the European Parliament on a European Community biodiversity strategy', COM (1998) 42

EC (2001) 'Communication from the Commission to the Council and the European Parliament on the biodiversity action plan for economic and development co-operation', COM (2001) 162

EC (2003) 'Communication from the Commission to the Council and the European Parliament on implementation by the EC of the "Bonn Guidelines" on access to genetic resources and benefit-sharing under the Convention on Biological Diversity', COM (2003) 821 final

EC-CHM (Clearing-House Mechanism) (2012) 'Home site', http://biodiversity-chm.eea.europa.eu, accessed 12 November 2012

Edwards, P.J. and Abivardi, C. (1998) 'The value of biodiversity: where ecology and economy blend', *Biodiversity Conservation*, vol. 83, no. 3, pp.239–246

Eichler, J. (2001) 'Biotechnological uses of archaeal extremozymes', *Biotechnology Advances*, vol. 19, pp.261–271

Elan (2011) *Annual Report*, http://ir.elan.com/phoenix.zhtml?c=88326&p=irol-reportsannual, accessed 6 November 2012

Elian, G. (1979) *The principle of sovereignty over natural resources*, Sijthoff & Noordhoff, Alphen ann den Rijn, the Netherlands

Encode Project Consortium (2012) 'An integrated encyclopedia of DNA elements in the human genome', *Nature*, vol. 489, pp.57–74

Engle, C.R. (2009) 'Mariculture, economic and social impacts', in J.H. Steele, K.K. Turekian and S.A. Thorpe (eds) *Encyclopedia of ocean sciences*, Elsevier, Amsterdam

Estée Lauder (2012) *Annual Report*, http://investors.elcompanies.com/phoenix. zhtml?c=109458&p=irol-reports, accessed 6 November 2012

European Bioinformatics Institute (2012) 'UniProtKB/TrEMBL protein database release 2011_01 statistics', www.ebi.ac.uk/uniprot/TrEMBLstats, accessed 15 November 2012

European Commission (2008) *The European Union's biodiversity action plan. Halting the loss of biodiversity by 2010 – and beyond*, Office for Official Publications of the European Communities, Luxembourg

EvaluatePharma (2012) 'Australian and Swedish experts hunt bacteria blockers', www. evaluatepharma.com/Universal/View.aspx?type=Story&id=162657, accessed 12 November 2012

FAO (Food and Agriculture Organization) (1983) 'International Undertaking on Plant Genetic Resources', FAO Conference Resolution 8/83

FAO (1989) 'Agreed interpretation of the International Undertaking', FAO conference resolution 4/89

FAO (1991) FAO conference resolution 3/91

FAO (2004) *State of world aquaculture*, FAO, Rome

Farrier, D. and Tucker, L. (2001) 'Access to marine bioresources: hitching the conservation cart to the bioprospecting horse', *Ocean Development and International Law*, vol. 32, pp.213–239

FDA (Food and Drug Administration) (2010) *Environmental assessment for AquAdvantage® salmon*, FDA, Silver Spring, USA

Federal Ministry for the Environment, Nature Conservation and Nuclear Safety (BMU) (2007) *National Strategy on Biological Diversity*, Bonifatius, Paderborn, Germany

Fenchel, T. and Finlay, B.J. (2004) 'The ubiquity of small species: patterns of local and global diversity', *BioScience*, vol. 54, no. 8, pp.777–784

Fenical, W. (2006) 'Marine pharmaceuticals. Past, present, and future', *Oceanography*, vol. 19, no. 2, pp.110–119

Flitner, M. (1998) 'Biodiversity: of local commons and global commodities', in M. Goldmann (ed.) *Privatizing nature. Political struggles for the global commons*, Pluto Press, London

Fore*sight* Marine Panel (2005) *A study into the prospects for marine biotechnology development in the United Kingdom*, vol. 2, Institute of Marine Engineering, London

Francioni, F. (2006) 'International law for biotechnology: basic principles', in F. Francioni and T. Scovazzi (eds) *Biotechnology and international law*, Hart Publishing, Oxford

Freestone, D. (1995) 'The conservation of marine ecosystems under international law', in C. Redgwell and M. Bowman (eds) *International law and the conservation of biological diversity*, Kluwer Law International, London

Frenz, J.L., Kohl, A.C. and Kerr, R.G. (2004) 'Marine natural products as therapeutic agents: part 2', *Experts Opinions in Therapeutic Patents*, vol. 14, no. 1, pp.17–33

Füllbeck, M., Michalsky, E., Dunkel, M. and Preissner, R. (2006) 'Natural products: sources and databases', *Natural Product Reports*, vol. 23, pp.347–356

Galperin, M.Y. and Cochrane, G.R. (2011) 'The 2011 Nucleic Acids Research database issue and the online molecular biology database collection', *Nucleic Acids Research*, vol. 39, pp.D1–D6

Garforth, K. and Medaglia, J.C. (2006) 'Factors contributing to legal reform for the development and implementation of measures on access to genetic resources and benefit sharing', in T. McInerney (ed.) *Searching for success. Narrative accounts of legal reform in developing and transition countries*, International Development Law Organization, Rome

Garforth, K., Noriega, I.L., Medaglia, J.C., Nnadozie, K. and Nemogá, G.R. (2005) *Overview of the national and regional implementation of access to genetic resources and benefit sharing measures*, CISDL, Montreal

Garrity, G.M., Thompson, L.M., Ussery, D.W., Paskin, N., Baker, D., Desmeth, P., Schindel, D.E. and Ong, P.S. (2009) 'Studies on monitoring and tracking genetic resources', CBD WG-ABS information document UNEP/CBD/WG-ABS/7/INF/2

Gavouneli, M. (2007) *Functional jurisdiction in the law of the sea*, Martinus Nijhoff Publishers, Leiden

GBIF (Global Biodiversity Information Facility) (2007) *The GBIF data portal*, GBIF, Copenhagen

GBIF (2012a) 'Welcome to the GBIF data portal', http:// data.gbif.org/welcome.htm, accessed 16 November 2012

GBIF (2012b) 'Standard and tools', www.gbif.org/informatics/standards-and-tools, accessed 15 November 2012

GBIF (2012c) 'GBIF Data Sharing Agreement', http://data.gbif.org/tutorial/datasharingagreement, accessed 15 November 2012

GBIF (2012d) 'GBIF Data Use Agreement', http:// data.gbif.org/tutorial/datauseagreement, accessed 15 November 2012

GBIF Secretariat (2010) *Memorandum of understanding for the Global Biodiversity Information Facility*, GBIF Secretariat, Copenhagen

GEF (Global Environment Facility) (2008) *Instrument for the establishment of the restructured Global Environment Facility*, GEF, Washington DC

GEF (2010) 'System for Transparent Allocation of Resources (STAR)', GEF document GEF/P.3

German Advisory Council on Global Change (2010) *Future bioenergy and sustainable land use*, Earthscan, London

Glowka, L. (1996) 'The deepest of ironies: genetic resources, marine scientific research, and the area', *Ocean Yearbook*, vol. 12, pp.154–178

Glowka, L. (1997a) 'Emerging legislative approaches to implement article 15 of the Convention on Biological Diversity', *RECIEL*, vol. 6, no. 3, pp.249–262

Glowka, L. (1997b) 'Legal and institutional considerations for states providing genetic resources', in J. Mugabe, C.V. Barber, G. Henne, L. Glowka and A. la Viña (eds) *Access to genetic resources: strategies for sharing benefits*, ACTS, Nairobi

Glowka, L. (1998) *A guide to designing legal frameworks to determine access to genetic resources*, IUCN, Gland, Switzerland

Glowka, L. (1999) 'Genetic resources, marine scientific research and the international seabed area', *RECIEL*, vol. 8, no. 1, pp.56–66

Glowka, L., Burhenne-Guilmin, F. and Synge, H. (1994) *A guide to the Convention on Biological Diversity*, IUCN, Gland, Switzerland

Gong, Z., Wan, H., Tay, T.L., Wang, H., Chen, M. and Yan, T. (2003) 'Development of transgenic fish for ornamental and bioreactor by strong expression of fluorescent proteins in the skeletal muscle', *Biochemical and Biophysical Research Communications*, vol. 308, pp.58–63

Goodman, N. (2002) 'Biological data becomes computer literate: new advances in bioinformatics', *Current Opinion in Biotechnology*, vol. 13, pp.68–71

Görg, C. and Brand, U. (2000) 'Global environmental politics and competition between nation-states: on the regulation of biological diversity', *Review of International Political Economy*, vol. 7, no. 3, pp.371–398

Gorina-Ysern, M. (1998) 'Marine scientific research activities as the legal basis for intellectual property claims?', *Marine Policy*, vol. 22, no. 4–5, pp.337–357

Gorina-Ysern, M. (2003) *Marine scientific research*, Transnational Publishers, Ardsley, New York, USA

Gorina-Ysern, M. and Jones, J.H. (2006) 'International law of the sea, access and benefit sharing agreements, and the use of biotechnology in the development, patenting and commercialization of marine natural products as therapeutic agents', *Ocean Yearbook*, vol. 20, pp.221–281

Governing Body (2006) 'First session of the Governing Body of the International Treaty on Plant Genetic Resources for Food and Agriculture', Governing Body Report IT/GB-1/06/Report

Governing Body (2011a) 'Report on the implementation of the Multilateral System of Access and Benefit Sharing', Governing Body document IT/GB-4/11/12

Governing Body (2011b) 'Reviews and assessments under the Multilateral System, and of the implementation and operation of the standard material transfer agreement', Governing Body document IT/GB-4/11/13

Graf Vitzthum, W. (2006) 'Maritimes Aquitorium und Anschlusszone', in W. Graf Vitzthum (ed) *Handbuch des Seerechts*, Verlag C.H. Beck, Munich

Grajal, A. (1999) 'Biodiversity and the nation state: regulating access to genetic resources limits biodiversity research in developing countries', *Conservation Biology*, vol. 13, no. 1, pp.6–10

Grassle, J.F. (1989) 'Species diversity in deep-sea communities', *Tree*, vol. 4, no. 1, pp.12–15

Grassle, J.F. (2000) 'The Ocean Biogeographic Information System (OBIS): an on-line, worldwide atlas for accessing, modeling and mapping marine biological data in a multidimensional geographic context', *Oceanography*, vol. 13, no. 3, pp.5–7

Grassle, J.F. and Maciolek, N.J. (1992) 'Deep-sea species richness: regional and local diversity estimates from quantitative bottom samples', *The American Naturalist*, vol. 139, no. 2, pp.313–341

Grassle, J.F. and Stocks, K.I. (1999) 'A global ocean biogeographic information system (OBIS) for the Census of Marine Life', *Oceanography*, vol. 12, no. 3, pp.12–14

Gray, J.S. (1997) 'Marine biodiversity: patterns, threats and conservation needs', *Biodiversity and Conservation*, vol. 6, pp.153–175

Greenland Government (2012) 'Biological resources', http:// uk.nanoq.gl/Emner/ About/ Resources_and_industry/Biological_resources.aspx, accessed 12 November 2012

Greer, D. and Harvey, B. (2004) *Blue genes: sharing and conserving the world's aquatic biodiversity*, Earthscan, London

Greiber, T. (2011) *Access and benefit sharing in relation to marine genetic resources from areas beyond national jurisdiction. A possible way forward*, Bundesamt für Naturschutz, Bonn

Greiber, T., Moreno, S.P., Åhrén, M., Carrasco, J.N., Kamau, E.C., Medaglia, J.C. and Perron-Welch, M.J.O.F. (2012) *An explanatory guide to the Nagoya Protocol on Access and Benefit-sharing*, IUCN, Gland, Switzerland

Groombridge, B. and Jenkins, M.D. (2002) *World atlas of biodiversity*, University of California Press, Berkeley, CA

GSK (GlaxoSmithKline) (2011) *Annual Report*, www.gsk.com/investors/annual-reports. html, accessed 6 November 2012

Guilford-Blake, R. and Strickland, D. (2008) *Guide to biotechnology*, Biotechnology Industry Organization, Washington DC

Guinotte, J.M., Bartley, J.D., Iqbal, A., Fautin, D.G. and Buddemeier, R.W. (2006) 'Modeling habitat distribution from organism occurrences and environmental data: case study using anemonefishes and their sea anemone hosts', *Marine Ecology Progress Series*, vol. 316, pp.269–283

Haefner, B. (2003) 'Drugs from the deep: marine natural products as drug candidates', *DDT*, vol. 8, no. 12, pp.536–544

Hafner, G. (2006) 'Meeresumweltschutz, Meeresforschung und Technologietransfer', in W. Graf Vitzthum (ed.) *Handbuch des Seerechts*, Verlag C.H. Beck, Munich

Halle, S. (2009) *From conflict to peacebuilding. The role of natural resources and the environment*, UNEP, Nairobi

Halvorson, H.O. and Quezada, F. (2009) 'Marine biotechnology', in J.H. Steele, K.K. Turekian and S.A. Thorpe (eds) *Encyclopedia of ocean sciences*, Academic Press, San Diego, CA

Handl, G. (1978) 'The principle of "equitable use" as applied to internationally shared natural resources: its role in resolving potential international disputes over transfrontier pollution', *Revue belge do droit international*, vol. 14, no. 40, pp.40–64

Harper, M.K., Bugni, T.S., Copp, B.R., James, R.D., Lindsay, B.S., Richardson, A.D., Schnabel, P.C., Tasdemir, D., VanWagoner, R.M., Verbitski, S.M. and Ireland, C.M. (2001) 'Introduction to the chemical ecology of marine natural products', in I. McClintock, J.B. Baker and B. James (eds) *Marine chemical ecology*, CRC Press, Boca Raton, FL

Head, I.M., Martin Jones, D. and Röling, W.F.M. (2006) 'Marine microorganisms make a meal of oil', *Nature Reviews Microbiology*, vol. 4, pp.173–182

Heckrodt, T.J. and Mulzner, J. (2005) 'Marine natural products from *Pseudopterogorgia elisabethae*: structures, biosynthesis, pharmacology, and total synthesis', *Topics in Current Chemistry*, vol. 244, pp.1–41

Henne, G. (1997) 'Mutually agreed terms in the CBD: requirements under public international law', in J. Mugabe, C.V. Barber, G. Henne, L. Glowka and A. la Viña (eds) *Access to genetic resources: strategies for sharing benefits*, ACTS Press, Nairobi

Henne, G. (1998) *Genetische Vielfalt als Ressource*, Nomos, Baden-Baden, Germany

Hew, C.L. and Fletcher, G.L. (2001) 'The role of aquatic biotechnology in aquaculture', *Aquaculture*, vol. 197, pp.191–204

Hoagland, P., Jacoby, J. and Schumacher, M.E. (2009) 'Law of the sea', in J.H. Steele, S.A. Thorpe and K.K. Turekian (eds) *Encyclopedia of ocean sciences*, Academic Press, San Diego, CA

Holm-Müller, K., Richerzhagen, C. and Täuber, S. (2005) *Users of genetic resources in Germany*, BfN, Bonn

Horrocks, L.A. and Yeo, Y.K. (1999) 'Health benefits of docosahexaenoic acid (DHA)', *Pharmacological Research*, vol. 40, no. 3, pp.211–225

Hortal, J., Lobo, J.M. and Jiménez-Valverde, A. (2007) 'Limitations of biodiversity databases: case study on seed-plant diversity in Tenerife, Canary Islands', *Conservation Biology*, vol. 21, no. 3, pp.853–863

Howell, D.J. and Evans, S.M. (2009) 'Antifouling materials', in J.H. Steele, K.K. Turekian, and S.A. Thorpe (eds) *Encyclopedia of ocean sciences*, Elsevier, Amsterdam

Hughes Martiny, J.B., Bohannan, B.J.M., Brown, J.H., Colwell, R.K., Fuhrmann, J.A., Green, J.L., Horner-Devine, M.C., Kane, M., Adams Krumins, J., Kuske, C.R., Morin, P.J., Naeem, S., Øvreås, L., Reysenbach, A.-L., Smith, V.H. and Staley, J.T. (2006) 'Microbial biogeography: putting microorganisms on the map', *Nature Reviews*, vol. 4, pp.102–112

Hunt, B. and Vincent, A.C.J. (2006) 'Scale and sustainability of marine bioprospecting for pharmaceuticals', *Ambio*, vol. 35, no. 2, pp.57–64

ILC (International Law Commission) (2012) 'Shared natural resources (oil and gas)', http:// untreaty.un.org/ilc/guide/8_6.htm, accessed 13 November 2012

Informa (2012) 'Home site', www.informa.com, accessed 14 November 2012

IOC (Intergovernmental Oceanographic Commission) (2005), *IOC criteria and guidelines on the transfer of marine technology*, UNESCO, Paris

IOC (2009) 'The Ocean Biogeographic Information System (OBIS)', IOC Assembly resolution XXV-4

IOC (2012) 'National legislations', http://ioc-unesco.org/index.php?option=com_content &view=category&layout=blog&id=45&Itemid=100026, accessed 12 November 2012

IOC of UNESCO (2012) 'The Ocean Biogeographic Information System', www.iobis. org, accessed 15 November 2012

IUCN (International Union for the Conservation of Nature) and BfN (Bundesamt für Naturschutz) (2011) *Seminar on conservation and sustainable use of marine biodiversity beyond national jurisdiction*, Bundesamt für Naturschutz, Bonn

Jacobson, A.H. and Willingham, G.L. (2000) 'Sea-nine antifoulant: an environmentally acceptable alternative to organotin antifoulants', *The Science of the Total Environment*, vol. 258, pp.103–110

Jain, E., Bairoch, A., Duvaud, S., Phan, I., Redaschi, N., Suzek, B.E., Martin, M.J., McGarvey, P. and Gasteiger, E. (2009) 'Infrastructure for the life sciences: design and implementation of the UniProt web site', *BMC Bioinformatics*, vol. 10, pp.136–155

JCVI (J. Craig Venter Institute) (2007) 'Ocean Metagenomics', *PLoS Biology*, vol. 5, no. 3

JCVI (2012) 'Sorcerer II Expedition', www.sorcerer2expedition.org/version1/HTML/ main.htm, accessed 12 November 2012

Jeffery, M.I. (2002) 'Bioprospecting: access to genetic resources and benefit-sharing under the Convention on Biodiversity and the Bonn Guidelines', *Singapore Journal of International and Comparative Law*, vol. 6, pp.747–808

Jones, J.S. (2004) 'Regulating access to biological and genetic resources in Australia: a case study of bioprospecting in Queensland', in N.P. Stoianoff (ed.) *Accessing biological resources: complying with the Convention on Biological Diversity*, Kluwer Law International, the Netherlands

Joyner, C.C. (1995) 'Biodiversity in the marine environment: resource implications for the law of the sea', *Vanderbilt Journal of Transnational Law*, vol. 28, no. 4, pp.635–697

Kaczorek, E., Urbanowicz, M. and Olszanowski, A. (2010) 'The influence of surfactants on cell surface properties of *Aeromonas hydrophila* during diesel oil biodegradation', *Colloids and Surfaces B: Biointerfaces*, vol. 81, pp.363–368

Kaczorowska, A. (2010) *Public international law*, Routledge, Abingdon

Kamau, E.C. (2011a) 'Common pools of genetic resources – a potential approach in resolving inefficiency and injustice in ABS', in U. Feit and H. Korn (eds) *Genetic resources, traditional knowledge, and the law. Solutions for access and benefit sharing*, BfN, Bad Godesberg, Germany

Kamau, E.C. (2011b) 'The Multilateral System of the FAO Treaty: lessons for ABS for genetic diversity of global importance', *Revista International de Direito e Cidadania*

Kamau, E.C., Fedder, B. and Winter, G. (2010) 'The Nagoya Protocol on Access to Genetic Resources and Benefit Sharing: what is new and what are the implications for provider and user countries and the scientific community', *Law, Environment and Development Journal*, vol. 6, no. 3, pp.246–262

Karentz, D. (2001) 'Chemical defenses of marine organisms against solar radiation exposure: UV-absorbing mycosporine-like amino acids and scytonemin', in J.B. McClintock, B.J. Baker and B. James (eds) *Marine chemical ecology*, CRC Press, Boca Raton, FL

Kennedy, J. (2008) 'Mutasynthesis, chemobiosynthesis, and back to semi-synthesis: combining synthetic chemistry and biosynthetic engineering for diversifying natural products', *Natural Products Report*, vol. 25, pp.25–34

KGS (Kansas Geological Survey) (2012) 'Instructions and help files for KGSMapper', http:// drysdale.kgs.ku.edu/website/Specimen_Mapper/Mapper2Help.cfm, accessed 15 November 2012

Kiely, T., Donaldson, D. and Grube, A. (2004) *Pesticides industry sales and usage*, Environmental Protection Agency, Washington DC

Kijjoa, A. and Sawangwong, P. (2004) 'Drugs and cosmetics from the sea', *Marine Drugs*, vol. 2, pp.73–82

Kim, S.-K. and Mendis, E. (2006) 'Bioactive compounds from marine processing byproducts – a review', *Food Research International*, vol. 39, pp.383–393

Kimata, N., Nishino, T., Suzuki, S. and Kogure, K. (2004) '*Pseudomonas aeruginosa* isolated from marine environments in Tokyo Bay', *Microbial Ecology*, vol. 47, pp.41–47

Kirchner, A. (2010) 'Bioprospecting, marine scientific research and the patentability of genetic resources', in N.A. Martínez (ed.) *Serving the rule of international maritime law*, Routledge, Abingdon, UK

Klein, N. (2005) *Dispute settlement in the UN Convention on the Law of the Sea*, Cambridge University Press, Cambridge

Klein, D. and Düppen, A. (2008) 'ABS: die dritte Säule des Übereinkommens über die biologische Vielfalt', *Natur und Landschaft*, vol. 2, pp.42–46

Kloppenburg, J.R. (1988) *First the seed. The political economy of plant biotechnology 1492–2000*, Cambridge University Press, Cambridge

Köhler, J. (2004) 'Integration of life science databases', *DDT:Biosilico*, vol. 2, no. 2, pp.61–69

Korn, H., Friedrich, S. and Feit, U. (2003) *Deep sea genetic resources in the context of the Convention of Biological Diversity and the United Nations Convention on the Law of the Sea*, Bundesamt für Naturschutz, Bonn, Germany

Kornprobst, J.-M. (2010) *Encyclopedia of marine natural products*, Wiley-Blackwell, Weinheim

Kriwoken, L.K. (1996) 'Australian biodiversity and marine protected areas', *Ocean and Coastal Management*, vol. 33, no1–3, pp.113–132

Kuhlmann, J. (1997) 'Drug research: from the idea to the product', *International Journal of Clinical Pharmacology and Therapeutics*, vol. 35, pp.541–552

Lagoni, R. and Proelß, A. (2006) 'Festlandsockel und ausschließliche Wirtschaftszone', in W. Graf Vitzthum (ed) *Handbuch des Seerechts*, Verlag C.H. Beck, Munich

Laird, S., Monagle, C. and Johnston, S. (2008) *Queensland Biodiscovery Collaboration. The Griffith University AstraZeneca partnership for natural product discovery. An access and benefit case study*, UNU-IAS, Yokohama

Lambshead, P.J.D. (1993) 'Recent developments in marine benthic biodiversity research', *Oceanus*, vol. 19, no. 6, pp.5–24

Lane, M.A. and Edwards, J.L. (2007) 'The Global Biodiversity Information Facility (GBIF)', in G.B. Curry and C.J. Humphries (eds) *Biodiversity databases: techniques, politics, and applications*, CRC Press, Boca Raton, FL

Lasserre, P. (1994) 'The role of biodiversity in marine ecosystems', in O.T. Solbrig, H.M. van Emden and P.G.W.J. van Oordt (eds) *Biodiversity and global change*, CABI Publishing, Wallingford, UK

Leary, D. (2006) *International law and the genetic resources of the deep sea*, Nijhoff, Leiden

Leary, D. (2012) 'Moving the marine genetic resources debate forward: some reflections', *The International Journal of Marine and Coastal Law*, vol. 27, pp.435–448

Leary, D., Vierros, M., Hamon, G., Arico, S. and Monagle, C. (2009) 'Marine genetic resources: a review of scientific and commercial interest', *Marine Policy*, vol. 33, pp.183–194

Lesser, W. (1998) *Sustainable use of genetic resources under the Convention on Biological Diversity: exploring access and benefit sharing issues*, CABI, London

Levy, S., Sutton, G., Ng, P.C., Feuk, L., Halpern, A.L., Walenz, B.P., Axelrod, N., Huang, J., Kirkness, E.F., Denisov, G., Lin, Y., MacDonald, J.R., Wing Chun Pang, A., Shago, M., Stockwell, T.B., Tsiamouri, A., Bafna, V., Bansal, V., Kravitz, S.A., Busam, D.A., Beeson, K.Y., McIntosh, T.C., Remington, K.A., Abrill, J.F., Gill, J., Borman, J., Rogers, Y.-H., Frazier, M.E., Sherer, S.W., Strausberg, R.L. and Venter, J.C. (2007) 'The diploid genome sequence of an individual human', *PLoS Biology*, vol. 5, no. 10, pp.2113–2144

Li, J.W.-H. and Vederas, J.C. (2009) 'Drug discovery and natural products: end of an era or an endless frontier?', *Science*, vol. 325, pp.161–165

Liles, G. (1996) 'Gambling on marine biotechnology', *BioScience*, vol. 46, no. 4, pp.250–253

Lindeskog, S. (2004) 'Bioprospecting, access and benefit sharing and traditional knowledge', in H.H. Lidgård (ed.) *Transferring technology to developing countries*, University of Lund, Lund, Sweden

Lintner, K., Mas-Chamberlin, C., Mondon, P., Peschard, O. and Lamy, L. (2009) 'Cosmeceuticals and active ingredients', *Clinics in Dermatology*, vol. 27, pp.461–468

Llewellyn, L.E. and Burnell, J.N. (2000) 'Marine organisms as sources of C_4-weed-specific herbicides', *Pesticide Outlook*, pp.64–67

Lochen, T. (2007) *Die völkerrechtlichen Regelungen über den Zugang zu genetischen Ressourcen*, Mohr Siebeck, Tübingen, Germany

Loup, W.D., Farnsworth, N.R., Soejarto, D.D. and Quinn, M.L. (1985) 'NAPRALERT: Computer handling of natural product research data', *Journal of Chemical Informatics and Computer Sciences*, vol. 25, pp.99–103

Lupin (2012) 'Products', www.lupinpharmaceuticals.com/specialty.htm, accessed 6 November 2012

Maclean, N. (1998) 'Regulation and exploitation of transgenes in fish', *Mutation Research*, vol. 399, pp.255–266

Mahmoudi, S. (1999) 'Common heritage of mankind, common concern of humanity', in J.-P. Beurier, A. Kiss and S. Mahmoudi (eds) *New technologies and law of the marine environment*, Kluwer Law International, London

Malanczuk, P. (2002) *Akehurst's modern introduction into international law*, Routledge, London

Marston, G. (1989) 'Maritime jurisdiction', in R. Bernhardt (ed.) *Encyclopedia of public international law*, North-Holland Publ. Co., Amsterdam

Martek (2009) *Annual Report*, files.shareholder.com/downloads/MATK/1008109369x0x 346470/5a6fab44-d803-4f9b-9401-3c5c58440875/MATK_2009_Annual.pdf, accessed 6 November 2012

Mattila, P., Korpela, J., Tenkanen, T. and Pitkänen, K. (1991) 'Fidelity of DNA synthesis by the *Thermococcus litoralis* DNA polymerase – an extremely heat stable enzyme with proofreading activity', *Nucleic Acids Research*, vol. 19, no. 18, pp.4967–4973

Matz, N. (2002) 'The interaction between the Convention on Biological Diversity and the UN Convention on the Law of the Sea', in P. Ehlers, E. Mann-Borgese and R. Wolfrum (eds) *Marine issues. From a scientific, political, and legal perspective*, Kluwer Law International, London

May, R.M. (1992) 'Bottoms up for the oceans', *Nature*, vol. 357, pp.278–279

May, R.M. (1994) 'Biological diversity: differences between land and sea', *Philosophical Transactions of the Royal Society of London B*, vol. 343, pp.105–111

McGraw, D.M. (2002) 'The CBD – key characteristics and implications for implementation', *RECIEL*, vol. 11, no. 1, pp.17–28

McLaughlin, R.J. (2003) 'Foreign access to shared marine genetic materials: management options for a quasi-fugacious resource', *Ocean Development and International Law*, vol. 34, pp.297–348

Medaglia, J.C. (2004) *A comparative analysis of the implementation of access and benefit sharing regulations in selected countries*, IUCN, Bonn

Medaglia, J.C. and Silva, C.I. (2007) *Addressing the problems of access: protecting sources, while giving users certainty*, IUCN, Gland, Switzerland

Meisel, M. (2009) 'Green is the new black: innovations abound in natural ingredients for household and personal care products', *Focus on Surfactants*, vol. 9, p.6

Melamed, P., Gong, Z., Fletcher, G. and Hew, C.L. (2002) 'The potential impact of modern biotechnology on fish aquaculture', *Aquaculture*, vol. 204, pp.255–269

Meng, P.-J., Wang, J.-T., Liu, L.-L., Chen, M.-H. and Hung, T.-C. (2005) 'Toxicity and bioaccumulation of tributyltin and triphenyltin on oysters and rock shells collected from Taiwan mariculture area', *Science of the Total Environment*, vol. 349, pp.140–149

Merck (2012) 'Search Guide. The Merck Index 14th Edition CD', http://www.merckbooks. com/mindex/pdf/Web_User_Help_Guide.pdf, accessed 14 November 2012

Mgbeoji, I. (2003) 'Beyond rhetoric: state sovereignty, common concern, and the inapplicability of the common heritage concept to plant genetic resources', *Leiden Journal of International Law*, vol. 16, pp.821–837

Miller, J.S. (2007) 'Impact of the Convention on Biological Diversity: the lessons of ten years of experience with models for equitable sharing of benefits', in C.R. McManis (ed.) *Biodiversity and the law. Intellectual property, biotechnology and traditional knowledge*, Earthscan, London

Minh, C.V., Kiem, P.V. and Dang, N.H. (2005) 'Marine natural products and their potential application in the future', *AJSTD*, vol. 22, no. 4, pp.297–311

Molinski, T.F., Dalisay, D.S., Lievens, S.L. and Saludes, J.P. (2009) 'Drug development from marine natural products', *Nature Reviews*, vol. 8, pp.69–85

Moore, G. and Tymowski, W. (2005) *Explanatory guide to the International Treaty on Plant Genetic Resources for Food and Agriculture*, IUCN, Gland, Switzerland

Moran, K., King, S.R. and Carlson, T.J. (2001) 'Biodiversity prospecting: lessons and prospects', *Annual Review of Anthropology*, vol. 30, pp.505–526

Morse, D.E. (1986) 'Biotechnology in marine aquaculture', *Aquacultural Engineering*, vol. 5, pp.347–355

Mugabe, J., Barber, C.V., Henne, G., Glowka, L. and la Viña, A. (1997) 'Managing access to genetic resources', in J. Mugabe, C.V. Barber, G. Henne, L. Glowka and A. la Viña (eds) *Access to genetic resources. Emerging regimes to facilitate regulation and benefit-sharing*, ACTS Press, Nairobi

Mulligan, C.N. (2009) 'Recent advances in the environmental applications of biosurfactants', *Current Opinion in Colloid & Interface Science*, vol. 14, pp.372–378

Mutanda, T., Ramesh, D., Karthikeyan, S., Kumari, S., Anandraj, A. and Bux, F. (2011) 'Bioprospecting for hyper-lipid producing microalgal strains for sustainable biofuel production', *Bioresource Technology*, vol. 102, no. 1, pp.57–70

Myers, N. (1988) 'Threatened biotas: "hot spots" in tropical forests', *The Environmentalist*, vol. 8, no. 3, pp.187–208

Nanda, V.P. and Pring, G. (2003) *International environmental law for the 21st century*, Transnational Publishers, New York

NAPRALERT (2012) 'Home site', www.napralert.org/About.aspx, accessed 15 November 2012

National Resource Management Ministerial Council (2002) *Nationally consistent approach for access to and the utilisation of Australia's native genetic and biochemical resources*, DEWHA, Canberra

Naylor, L.H. (1999) 'Reporter gene technology: The future looks bright', *Biochemical Pharmacology*, vol. 58, pp.749–757

NCBI (National Center for Biotechnology Information) (2009) *BLAST Basic Local Alignment Search Tool*, NCBI, Bethesda, MD

NCBI (2012a) 'Entrez, the life sciences search engine', www.ncbi.nlm.nih.gov/sites/gquery, accessed 15 November 2012

NCBI (2012b) 'PubChem Compound database, CID108150, Source: DrugBank', http://pubchem.ncbi.nlm.nih.gov/summary/summary.cgi?cid=108150&loc=ec_rcs, accessed 15 November 2012

NCBI (2012c) 'PubChem Deposition Gateway', http://pubchem.ncbi.nlm.nih.gov/deposit/deposit.cgi, accessed 15 November 2012

NCBI (2012d) 'Copyright and disclaimers', www.ncbi.nlm.nih.gov/About/disclaimer.html, accessed 15 November 2012

NCBI (2012e) 'PubMed', www.ncbi.nlm.nih.gov/pubmed, accessed 15 November 2012

NCBI (2012f) 'Synthetic construct opAFP-GHc2', www.ncbi.nlm.nih.gov/nuccore/56691717, accessed 15 November 2012

NCBI (2012g) 'Protein', www.ncbi.nlm.nih.gov/protein, accessed 15 November 2012

New England Biolabs (2012) 'Polymerases and amplification', www.neb.com/nebecomm/products/category6.asp?#7, accessed 3 November 2012

Newman, D.J. and Cragg, G.M. (2004) 'Marine natural products and related compounds in clinical and advanced clinical trial', *Journal of Natural Products*, 67, pp.1216–1238

Newman, D.J. and Cragg, G.M. (2007) 'Natural products as sources of new drugs over the last 25 years', *Journal of Natural Products*, vol. 70, pp.461–471

Newman, D.J., Kilama, J., Bernstein, A. and Chivian, E. (2008) 'Medicines from nature', in E. Chivian and A. Bernstein (eds) *Sustaining life*, Oxford University Press, Oxford

NewsMedical (2012) 'Griffith University and Pfizer in new partnership to unlock nature's cure for infections', www.news-medical.net/news/2008/10/27/42203.aspx, accessed 12 November 2012

Nijman, J. and Nollkaemper, A. (2007) *New perspectives on the divide between national and international law*, Oxford University Press, Oxford

Nordquist, M.H. (2002) *United Nations Convention on the Law of the Sea, 1982: a commentary, Vol. 4*, Nijhoff, Dordrecht, the Netherlands

Normand, V. (2004) 'Access to genetic resources and the fair and equitable sharing of benefits arising out of their utilization: developments under the Convention on Biological Diversity', *Journal of International Biotechnology Law*, vol. 1, pp.131–141

Norse, E.A. (1993) *Global marine biological diversity*, Island Press, Washington DC

OBIS (Ocean Biogeographic Information System) (2011) 'Home site', v2.iobis.org, accessed 21 January 2011 (site does not exist anymore)

O'Connell, D.P. (1984) 'The theory of maritime jurisdiction', in I.A. Shearer (ed.) *The international law of the sea*, Clarendon Press, Oxford

O'Dor, R., Miloslavich, P. and Yarincik, K. (2010) 'Marine biodiversity and biogeography – regional comparisons of global issues, an introduction', *PLoS*, vol. 5, no. 8, pp.1–7

OECD (Organisation of Economic Co-operation and Development) (2002) *Frascati Manual. Proposed standard practice for surveys on research and experimental development*, OECD, Paris

OECD (2012) 'Statistical definition of biotechnology', www.oecd.org/sti/biotechnology policies/statisticaldefinitionofbiotechnology.htm, accessed 3 November 2012

Office of the Special Representative of the Secretary-General for the Law of the Sea (1985) *Law of the Sea Bulletin*, vol 5

Öhman, M.D.P. (2002) *Access to and intellectual property rights over genetic resources with a special focus on fair and equitable benefit sharing*, International Institute for Industrial Environment Economics, Lund, Sweden

Oli, K.P., Feyerabend, G.B. and Lassen, B. (2010) *Towards an access and benefit sharing framework agreement for the genetic resources and traditional knowledge of the Hindu Kush-Himalayan region*, ICIMOD, Kathmandu, Nepal

Olivera, B.M. (2000) 'ω-conotoxin MVIIA: from marine snail venom to analgesic drug', in N. Fusetani (ed) *Drugs from the sea*, Karger, Basel

Olivera, B.M. (2006) 'Conus snail venom peptides', in A. Kastin (ed.) *Handbook of biologically active peptides*, Elsevier, Amsterdam

O'Malley, M.A. (2007) 'The nineteenth century roots of "everything is everywhere"', *Nature Reviews*, vol. 5, pp.647–651

O'Neil, M.J., Heckelman, P.E., Koch, C.B. and Roman, K.J. (eds) (2006) *The Merck Index: an encyclopedia of chemicals, drugs, and biologicals*, Merck & Co, Whitehouse Station, USA

Onwuekwe, C.B. (2007) 'Ideology of the commons and property rights: Who owns plant genetic resources and the associated traditional knowledge', in P.W.B Phillips and C.B. Onwuekwe (eds) *'Accessing and sharing the benefits of the genomics revolution*, Springer, Dordrecht

Orrego Vicuña, F. (1989) *The exclusive economic zone. Regime and legal nature under international law*, Cambridge University Press, Cambridge

Ortholand, J.-Y. and Ganesan, A. (2004) 'Natural products and combinatorial chemistry: back to the future', *Current Opinion in Chemical Biology*, vol. 8, pp.271–280

Owens, D. and Chambers, F. (2004) 'A study into the legal framework for marine biotechnology development in the United Kingdom', The Institute of Marine Engineering, Science and Technology Report: FMP MBG – 01

Paradell-Trius, L. (2000) 'Principles of international environmental law: an overview', *RECIEL*, vol. 9, no. 2, pp.93–99

Park, J.B.K., Craggs, R.J. and Shilton, A.N. (2011) 'Wastewater treatment high rate algal ponds for biofuel production', *Bioresource Technology*, vol. 102, no. 1, pp.35–42

Parry, B. (2000) 'The fate of the collections: social justice and the annexation of plant genetic resources', in C. Zerner (ed.) *People, plants and justice: the politics of nature conservation*, Columbia University Press, New York

Pavoni, R. (2006) 'Biodiversity and biotechnology: consolidation and strains in the emerging international legal regimes', in F. Francioni and T. Scovazzi (eds) *Biotechnology and international law*, Hart Publishing, Oxford

Peng, J., Shen, X., El Sayed, K.A., Dunbar, D.C., Perry, T.L., Wilkins, S.P., Hamann, M.T., Bobzin, S., Huesing, J., Camp, R., Prinsen, M., Krupa, D. and Wideman, M.A. (2003) 'Marine natural products as prototype agrochemical agents', *Journal of Agricultural and Food Chemistry*, vol. 51, pp.2246–2254

Perez, A., Chin-Ta, C. and Afero, F. (2009) 'Belize-Guatemala territorial dispute and its implications for conservation', *Tropical Conservation Science*, vol. 2, no. 1, pp.11–24

PharmaMar (2009) *Annual Report*, www.pharmamar.com/investors.aspx, accessed 6 November 2012

PharmaMar (2012) 'Healthcare professionals', www.pharmamar.com/products-professionals.aspx, accessed 6 November 2012

Phillips, M. (2009) 'Mariculture overview', in J.H. Steele, K.K. Turekian and S.A. Thorpe (eds) *Encyclopedia of ocean sciences*, Elsevier, Amsterdam

Pisupati, B. (2007) *Access to genetic resources, benefit sharing and bioprospecting*, UNU-IAS, Yokohama

Pisupati, B. (2008) 'Access and benefit sharing (ABS): issues and policy options', *Asian Biotechnology and Development Review*, vol. 10, no. 3, pp.1–2

Pisupati, B., Leary, D. and Arico, S. (2008) 'Access and benefit sharing: issues related to marine genetic resources', *Asian Biotechnology and Development Review*, vol. 10, no. 3, pp.49–68

Pomponi, S.A., Baden, D.G. and Zohar, Y. (2007) 'Marine biotechnology: realizing the potential', *Marine Technology Society Journal*, vol. 41, no. 3, pp.24–31

Prince, R.C. (1997) 'Bioremediation of marine oil spills', *TIBTECH*, vol. 15, pp.158–160

Proelß, A. (2009) 'ABS in relation to marine GRs', in E.C. Kamau and G. Winter (eds) *Genetic resources, traditional knowledge, and the law. Solutions for access and benefit sharing*, Earthscan, London

Proksch, P., Edrada, R.A. and Ebel, R. (2002) 'Drugs from the sea – current status and microbiological implications', *Applied Microbiological Biotechnology*, vol. 59, pp.125–134

Radmer, R.J. (1996) 'Algal diversity and commercial algal products', *BioScience*, vol. 46, no. 4, pp.263–270

RAFI (Rural Advancement Foundation International) (1994) *Bioprospecting/biopiracy and indigenous peoples*, RAFI communiqué, Ottawa, Canada

Rainne, J. (2006) 'The work of the International Commission on shared natural resources: the pursuit of competence and relevance', *Nordic Journal of International Law*, vol. 75, pp.321–338

Raveendran, T.V. and Limna Mol, V.P. (2009) 'Natural product antifoulants', *Current Science*, vol. 97, no. 4, pp.508–520

Reaka-Kudla, M.L. (1997) 'The global biodiversity of coral reefs: a comparison with rain forests', in M.L. Reaka-Kudla, D.E. Wilson and E.O. Wilson (eds) *Biodiversity II: understanding and protecting our biological resources*, Joseph Henry Press, Washington D.C.

Rédei, G.P. (2008) *Encyclopedia of genetics, genomics, proteonomics, and informatics*, Springer, Berlin

Reid, W.V. (1994) 'Biodiversity prospecting: strategies for sharing benefits', in V. Sánchez and C. Juma (eds) *Biodiplomacy. Genetic resources and international relations*, ACTS Press, Nairobi

Reid, W.V., Laird, S.A., Gámez, R., Sittenfeld, A., Janzen, D.H., Gollin, M.A. and Juma, C. (1993) 'A new lease on life', in W.V. Reid, S.A. Laird, C. Meyer, R. Gámez, A. Sittenfeld, D.H. Janzen, M.A. Gollin and C. Juma (eds) *Biodiversity prospecting: using genetic resources for sustainable development*, World Resource Institute, Washington DC

Reid, W.V., Barber, C.V. and La Viña, A. (1995) 'Translating genetic resources rights into sustainable development: gene cooperatives, the biotrade and lessons from the Philippines', *Plant Genetic Resources Newsletter*, vol. 102, pp.1–17

Reimers, C.E., Tender, L.M., Fertig, S. and Wang, W. (2001) 'Harvesting energy from the marine sediment–water interface', *Environmental Science and Technology*, vol. 35, pp.192–195

Ridgeway, L. (2009) 'Marine genetic resources: outcomes of the United Nations Informal Consultative Process (ICP)', *The International Journal of Marine and Coastal Law*, vol. 24, pp.309–331

Rimmer, M. (2009) 'The Sorcerer II expedition: intellectual property and biodiscovery', *MqJICEL*, vol. 6, pp.147–187

Rishton, G.M. (2008) 'Natural products as a robust source of new drugs and drug leads: past success and present day issues', *The American Journal of Cardiology*, vol. 101, pp.43D–49D

Rittschof, D. (2001) 'Natural product antifoulants' in J.B. McClintock and B.J. Baker (eds) *Marine chemical ecology*, CRC Press, Boca Raton, FL

Romiti, M. and Cooper, C. (2012) *Entrez help*, www.ncbi.nlm.nih.gov/books/NBK3837, accessed 15 November 2012

Rosenberg, E. and Ron, E.Z. (2001) 'Bioemulsans: surface-active polysaccharide-containing complexes', in A. Steinbuchel (ed) *Biopolymers*, Springer, New York

Rosendal, K. (1994) 'Implications of the US "no" in Rio', in V. Sanchez and C. Juma (eds) *Biodiplomacy: genetic resources and international relations*, ACTS Press, Nairobi

Rosenthal, J.P. (1997) 'Equitable sharing of biodiversity benefits: agreements on genetic resources', Paper presented at the International Conference on Incentive Measures for the Conservation and the Sustainable Use of Biological Diversity, Cairns, Australia

Ruiz, M. (2007) 'Accounting for the scientific present, technological advances and genetic information in the negotiations of the ABS international regime', *Policy and Environmental Law Series*, vol. 19, pp.1–4

Sadoff, C., Greiber, T., Smith, M. and Bergkamp, G. (2008) *Share. Managing water across boundaries*, IUCN, Gland, Switzerland

Sands, P. (2003) *Principles of international environmental law*, Cambridge University Press, Cambridge

Satpute, S.K., Banat, I.M., Dhakephalkar, P.K., Banpurkar, A.G. and Chopade, B.A. (2010) 'Biosurfactants, bioemulsifiers, and exopolysaccharides from marine microorganisms', *Biotechnology Advances*, vol. 28, pp.436–450

Sattelle, D.B., Harrow, I.D., David, J.A., Pelhate, M. and Callec, J.J. (1985) 'Nereistoxin: actions on a cns acetylcholine receptor/ion channel in the cockroach *Periplaneta americana*', *Journal of Experimental Biology*, vol. 118, pp.37–52

Sayers, E.W., Barrett, T., Benson, D.A., Bolton, E., Bryant, S.H., Canese, K., Chetvernin, V., Church, D.M., DiCuccio, M., Federhen, S., Feolo, M., Fingerman, I.M., Geer, L.Y., Helmberg, W., Kapustin, Y., Landsman, D., Lipman, D.J., Lu, Z., Madden, T.L., Magej, T., Maglott, D.R., Marchler-Bauer, A., Miller, V., Mizrachi, I., Ostell, J., Panchenko, A., Phan, L., Pruitt, K.D., Schuler, G.D., Sequeira, E., Sherry, S.T., Shumway, M., Sirotkin, K., Slotta, D., Souvorov, A., Starchenko, G., Tatusova, T.A., Wagner, L., Wang, Y., Wilbur, W.J., Yaschenko, E. and Ye, J. (2011) 'Database resources of the National Center for Biotechnology Information', *Nucleic Acids Research* ,vol. 39, pp.D38–D51

SCBD (Secretariat of the Convention on Biological Diversity) (2008) *Access and benefit-sharing in practice: trends in partnerships across sectors*, Technical Series No. 38, Montreal

Schander C. and Willassen, E. (2005) 'What can biological barcoding do for marine biology?', *Marine Biology Research*, vol. 1, pp.79–83

Schermers, H. (2002) 'Different aspects of sovereignty', in G. Kreijen (ed.) *State, sovereignty, and international governance*, Oxford University Press, Oxford

Scheuer, P.J. (1990) 'Some marine ecological phenomena: chemical basis and biomedical potential', *Science*, vol. 248, pp.173–177

Schnoes, A.M., Brown, S.D., Dodevski, I. and Babbitt, P.C. (2009) 'Annotation error in public databases: misannotation of molecular function in enzyme superfamilies', *PLoS Computational Biology*, vol. 5, no. 12, pp.1–13

Schrijver, N.J. (1988) 'Permanent sovereignty over natural resources versus the common heritage of mankind: complementary or contradictory principles of international economic law?', in M. Denters, P. Peters and P.J.I.M. Waart (eds) *International law and development*, Nijhoff, Dordrecht

Schrijver, N.J. (1993) 'Sovereignty and the sharing of natural resources', in A.H. Westing (ed.) *Transfrontier reserves for peace and nature: A contribution to human security*, UNEP, Nairobi

Schrijver, N.J. (1997) *Sovereignty over natural resources. Balancing rights and duties*, Cambridge University Press, Cambridge

Schroeder, R.A. (2000) 'Beyond distributive justice: resource extraction and environmental justice in the tropics', in C. Zerner (ed.) *People, plants and justice: the politics of nature conservation*, Columbia University Press, New York

Sederma (2012) 'Products details: Venuceane', www.sederma.fr/home.aspx?view=dtl&d=content&s=111&r=178&p=1138&prodID=112, accessed 6 November 2012

Sennett, S.H. (2001) 'Marine chemical ecology: applications in marine biomedical prospecting', in I. McClintock, J.B. Baker and B. James (eds) *Marine chemical ecology*, CRC Press, Boca Raton, FL

Shapiro, J.A. (2009) 'Revisiting the central dogma in the 21st century', *Annals of the New York Academy of Science*, vol. 1178, pp.6–28

Sherman, K. (2005) 'The large marine ecosystem approach for assessment and management of ocean coastal waters', in T.M. Hennessey and J.G. Sutinen (eds) *Sustaining large marine ecosystems. The human dimension*, Elsevier, Amsterdam

Shimizu, Y. (2000) 'Microalgae as a drug source', in N. Fusetani (ed) *Drugs from the sea*, Karger, Basel

Shimomura, O., Johnson, F.H. and Saiga, Y. (2005) 'Extraction, purification and properties of aequorin, a bioluminescent protein from the luminous hydromedusan, *Aequorea*', *Journal of Cellular and Comparative Physiology*, vol. 59, no. 3, pp.223–229

Shiva, V. (1997) *Biopiracy. The plunder of nature and knowledge*, South End Press, Boston, MA

Sims, N. (2000) *Pearl Oyster Information Bulletin 14*, Secretariat of the Community, Noumea, New Caledonia

Singh, A., Singh Nigam, P. and Murphy, J.D. (2011) 'Mechanism and challenges in commercialisation of algal biofuels', *Bioresource Technology*, vol. 102, no. 1, pp.26–34

So, P.-W., Parkes, H.G. and Bell, J.D. (2007) 'Application of magnetic resonance methods to studies of gene therapy', *Progress in Nuclear Magnetic Resonance Spectroscopy*, vol. 51, pp.49–62

Soons, A.H. (1982) *Marine scientific research and the law of the sea*, Kluwer, Deventer, the Netherlands

Soplín, S.P. and Muller, M.R. (2009) 'The development of an international regime of access to genetic resources and fair and equitable benefit sharing in context of new technological developments', *Initiative for the prevention of biopiracy*, vol. 4, no. 10, pp.1–16

Spolaore, P., Joannis, Cassan, C., Duran, E. and Isambert, A. (2006) 'Commercial applications of microalgae', *Journal of Bioscience and Bioengineering*, vol. 101, no. 2, pp.87–96

Steele, J.H. (1985) 'A comparison of terrestrial and marine ecological systems', *Nature*, vol. 313, pp.355–358

Steele, J.H. (1991a) 'Marine ecosystem dynamics: comparison of scales', *Ecological Research*, vol. 6, pp.175–183

Steele, J.H. (1991b) 'Marine functional diversity', *BioScience*, vol. 41, no. 7, pp.470–474

Steglich, W., Fugmann, B. and Lang-Fugmann, S. (2000) *RÖMPP encyclopedia natural products*, Thieme, Stuttgart, Germany

Stemplowski, R. (2006) 'Indivisible sovereignty and the European Union', in D.J. Eaton (ed.) *The end of sovereignty?: a transatlantic perspective*, Lit Verlag, Münster, Germany

Stoll, P.-T. (2004) 'Genetische Ressourcen, Zugang und Vorteilshabe', in N. Wolff and W. Köck (eds) *10 Jahre Übereinkommen über die biologische Vielfalt*, Nomos, Baden-Baden, Germany

Stoll, P.-T. (2009) 'Access to GRs and benefit sharing – underlying concepts and the idea of justice', in E.C. Kamau and G. Winter (eds) *Genetic resources, traditional knowledge, and the law. Solutions for access and benefit sharing*, Earthscan, London

Strathmann, R.R. (1990) 'Why life histories evolve differently in the sea', *American Zoologist*, vol. 30, pp.197–207

Sumithiradevi, C. and Punithavalli, M. (2009) 'Detecting redundancy in biological databases – an efficient approach', *Global Journal of Computer Science and Technology*, pp.141–145

Suneetha, M.S. and Pisupati, B. (2009) *Benefit sharing in ABS: options and elaborations*, UNU-IAS, Yokohama

Svarstad, H. (1994) 'National sovereignty and genetic resources', in V. Sánchez and C. Juma (eds) *Biodiplomacy: genetic resources and international relations*, ACTS Press, Nairobi

Svarstad, H., Bugge, H.C. and Dhillion, S.S. (2000) 'From Norway to Novartis: cyclosporin from *Tolypocladium inflatum* in an open access bioprospecting regime', *Biodiversity and Conservation*, vol. 9, pp.1521–1541

Swiderska, K. (2001) 'Stakeholder participation in policy on access to genetic resources, traditional knowledge and benefit-sharing', *Biodiversity and Livelihood Issues*, vol. 4, pp.1–32

Swiss Institute of Bioinformatics (2012a) 'Life science directory', www.expasy.org/links.html, accessed 14 November 2012

Swiss Institute of Bioinformatics (2012b) 'UniProtKB/Swiss-Prot protein knowledgebase release 2011_01 statistics', www.expasy.org/sprot/relnotes/relstat.html, accessed 15 November 2012

Takebayashi, Y., Pourquier, P., Zimonjic, D.B., Nakayama, K., Emmert, S., Ueda, T., Urasaki, Y., Kanzaki, A., Akiyama, S.-I., Popescu, N., Kraemer, K.H. and Pommier, Y. (2001) 'Antiproliferative activity of ecteinascidin 743 is dependent upon transcription-coupled nucleotide-excision repair', *Nature Medicine*, vol. 7, no. 8, pp.961–966

Tangley, L. (1996) 'Ground rules emerge for marine bioprospectors', *BioScience*, vol. 46, no. 4, pp.245–249

Taylor and Francis Group (2012) 'Home site', www.taylorandfrancisgroup.com, accessed 14 November 2012

ten Kate, K. (1997) 'The common regime on access to genetic resources in the Andean Pact', *Biopolicy Journal*, vol. 2, no. 6, pp.1–25

ten Kate, K. and Laird, S.A. (2000a) 'Biodiversity and business: coming to terms with the "grand bargain"', *International Affairs*, vol. 76, pp.241–264

ten Kate, K. and Laird, S.A. (2000b) *The commercial use of biodiversity*, Earthscan, London

Tender, L.M., Reimers, C.E., Stecher, H.A., Holmes, D.E., Bond, D.R., Lowy, D.A., Pilobello, K., Fertig, S.J. and Lovley, D.R. (2002) 'Harnessing microbially generated power on the seafloor', *Nature Biotechnology*, vol. 20, pp.821–825

Texaco Overseas Petroleum Co. and California Asiatic Oil Co. v the Government of the Libyan Arab Republic (1977), 53 ILR 389

Thakur, N.L., Thakur, A.N. and Müller, W.E.G. (2005) 'Marine natural products in drug discovery', *Natural Products Radiance*, vol. 4, no. 6, pp.471–477

The Government of the State of Kuwait > The American Independent Oil Co (1982), 21 ILM 976

Thieme (2010) *100% RÖMPP. Der professionelle Zugriff auf das gesicherte Wissen der Chemie*, Thieme, Stuttgart

Thieme (2012) 'Georg Thieme Verlag', www.thieme.de/index.html, accessed 14 November 2012

Thieme Chemistry (2012a) 'RÖMPP Online', www.roempp.com/prod/, accessed 14 November 2012

Thieme Chemistry (2012b) 'Home', www.thieme-chemistry.de/en/home.html, accessed 14 November 2012

Thorne-Miller, B. (1999) *The living ocean*, Island Press, Washington DC

Tobin, P. (2009) 'Zeltia sees wider Yondelis use leading to profit', www.bloomberg.com/apps/news?pid=newsarchive&sid=aMNY6ExT_AzE, accessed 6 November 2012

Treves, T. (1999) 'Dispute-settlement clauses in the Law of the Sea Convention and their impact on the protection of the marine environment: some observations', *RECIEL*, vol. 8, no. 1, pp.6–9

Tsioumani, E. (2004) 'International Treaty on Plant Genetic Resources for Food and Agriculture: legal and policy questions from adoption to implementation', *Yearbook of International Environmental Law*, vol. 15, pp.119–144

Tully, S. (2003) 'The Bonn Guidelines on access to genetic resources and benefit sharing', *RECIEL*, vol. 12, no. 1, pp.84–98

Tvedt, M.W. (2006) 'Elements for legislation in user countries to meet the fair and equitable benefit sharing commitment', *Journal of World Intellectual Property*, vol. 9, no. 2, pp.189–212

Tvedt, M.W. (2012) 'A report from the first reflection meeting on the global multilateral benefit-sharing mechanism', CBD ICNP information document UNEP/CBD/ICNP/2/INF/2

Tvedt, M.W. and Young, T. (2007) *Beyond access: exploring implementation of the fair and equitable sharing commitment in the CBD*, IUCN, Gland, Switzerland

UN (2002) 'Draft plan of implementation of the World Summit on Sustainable Development', UN document A/CONF.199/L.1

UN (2006) 'Report of the Ad Hoc Open-ended Informal Working Group to study issues relating to the conservation and sustainable use of marine biological diversity beyond areas of national jurisdiction', UN document A/61/65

UN (2007) 'Oceans and the law of the sea', United Nations Report of the Secretary-General A/62/66

UN (2011) 'Oceans and the law of the sea', United Nations Report of the Secretary-General A/66/70/Add. 1

UN (2012) 'Letter dated 8 June 2012 from the Co-Chairs of the Ad Hoc Open-ended Informal Working Group to the President of the General Assembly', UN document A/67/95

UN Security Council (1973) 'On peace and security in Latin America' UN Security Council Resolution 330

UNCTAD (United Nations Conference on Trade and Development) (2004) *The biotechnology promise. Capacity building for participation of developing countries in the bioeconomy*, UNCTAD document UNCTAD/ITE/IPC/2004/2

UNDOALOS (United Nations Division for Ocean Affairs and the Law of the Sea) (2010) *Marine scientific research. A revised guide to the implementation of the relevant provisions of the United Nations Convention on the Law of the Sea*, UN, New York

UNDOALOS (2012) 'Maritime space: Maritime zones and maritime delimitation', www.un.org/Depts/los/LEGISLATIONANDTREATIES/regionslist.htm, accessed 12 November 2012

UNEP (United Nations Environment Programme) (2006) *Marine and coastal ecosystems and human well-being: a synthesis report based on the findings of the Millennium Ecosystem Assessment*, UNEP, Nairobi

UNESCO (United Nations Educational, Scientific and Cultural Organization) (1962) 'Economic development and conservation of natural resources, flora and fauna', UNESCO General Conference resolution No. 12 C/2.213

UNGA (United Nations General Assembly) (1952a) 'Integrated economic development and commercial agreements', UNGA resolution A/RES/523 (VI)

UNGA (1952b) 'Right to exploit freely natural wealth and resources', UNGA resolution A/RES/626 (VII)

UNGA (1954) 'Recommendations concerning international respect for the right of peoples and nations to self-determination', UNGA resolution A/RES/837 (IX)

UNGA (1962a) 'Permanent sovereignty over natural resources', UNGA resolution A/RES/1803 (XVII)

UNGA (1962b) 'Economic development and the conservation of nature', UNGA resolution A/RES/1831 (XVII)

UNGA (1967) 'Examination of the question of the reservation exclusively for peaceful purposes of the sea-bed and the ocean floor, and the subsoil thereof, underlying the high seas beyond the limits of present national jurisdiction, and the use of their resources in the interests of mankind' UNGA document A/C.1/PV.1515 and 1516

UNGA (1970a) 'Permanent sovereignty over natural resources of developing countries and expansion of domestic sources of accumulation for economic development', UNGA resolution A/RES/2692 (XXV)

UNGA (1970b) 'Declaration on principles of international law concerning friendly relations and co-operation among states in accordance with the Charter of the United Nations', UNGA resolution A/RES/2625 (XXV)

UNGA (1971) 'Development and environment', UNGA resolution A/RES/2849 (XXVI)

UNGA (1972) 'Permanent sovereignty over natural resources of developing countries', UNGA resolution A/RES/3016 (XXVII)

UNGA (1973) 'Co-operation in the field of the environment concerning natural resources shared by two or more states', UNGA resolution A/RES/3129 (XXVIII)

UNGA (1973) 'Permanent sovereignty over natural resources', UNGA resolution A/RES/3171 (XXVIII)

UNGA (1974a) 'Charter of economic rights and duties of states', UNGA resolution A/RES/3281 (XXIX)

UNGA (1974b) 'Declaration on the establishment of a new international economic order', UNGA resolution A/RES/3201 (S-VI)

UNGA (1974c) 'Programme of action on the establishment of a new international economic order', UNGA resolution A/RES/3202 (S-VI)

UNGA (1978) 'Co-operation in the field of the environment concerning natural resources shared by two or more states', UNGA resolution A/RES/33/87

UNGA (1979) 'Co-operation in the field of the environment concerning natural resources shared by two or more states', UNGA resolution A/RES/34/186

UNGA (1982) 'World charter for nature', UNGA resolution A/RES/37/7

UNGA (1988) 'Protection of global climate for present and future generations of mankind', UNGA resolution A/RES/43/53

UNGA (2000) 'Results of the review by the Commission on Sustainable Development of the sectoral theme of "Oceans and seas": international coordination and cooperation', UNGA resolution A/RES/54/33

UNGA (2005) 'Convention on Biological Diversity', UNGA Resolution A/RES/59/236

UNGA (2009) 'Permanent sovereignty of the Palestinian people in the occupied Palestinian territory, including East Jerusalem, and of the Arab population in the occupied Syrian Golan over their natural resources', UNGA resolution A/RES/63/201

UniProt Consortium (2010) 'The Universal Protein Resource (UniProt) in 2010', *Nucleic Acids Research*, vol. 38, pp. D142–D148

UniProt Consortium (2012a) 'Reorganizing the protein space at the Universal Protein Resource (UniProt)', *Nucleic Acids Research*, vol. 40, pp.D71–D75, www.uniprot.org/uniprot/Q51334, accessed 15 November 2012

UniProt Consortium (2012b) 'License and disclaimer', www.uniprot.org/help/license, accessed 15 November 2012

United Kingdom of Great Britain and Northern Ireland v Iceland, ICJ Reports 1974

University of Illinois at Chicago (2012) 'Home site', www.uic.edu/uic/index.shtml, accessed 15 November 2012

UNU-IAS (United Nations University – Institute of Advanced Studies) (2008) *Access to genetic resources in Africa. Analysing ABS policy development in four African countries*, UNU-IAS, Yokohama, Japan

van den Hove, S. and Moreau, V. (2007) *Deep-sea biodiversity and ecosystems: a scoping report on their socio-economy, management and governance*, UNEP-WCMC Biodiversity Series, No. 28, Cambridge

Verhoosel, G. (1998) 'Prospecting for marine and coastal biodiversity: international law in deep water', *The International Journal of Marine and Coastal Law*, vol. 13, no. 1, pp.91–104

Vernooy, R. Haribabu, E., Muller, M.R., Vogel, J.H., Hebert, P.D.N., Schindel, D.E., Shimura, J. and Singer, G.A.C. (2010) 'Barcoding life to conserve biological diversity: beyond the taxonomic imperative', *PLoS Biology*, vol. 8, no. 7, pp.1–5

Vogel, J.H. (1997) 'The successful use of economic instruments to foster sustainable use of biodiversity: six case studies from Latin America and the Caribbean', *Biopolicy Journal*, vol. 2, no. 1

Vogel, J.H. (2000) *The Biodiversity Cartel. Transforming traditional knowledge into trade secrets*, CARE, Quito

Vogel, J.H. (2007a) 'Reflecting financial and other incentives of the TMOIFGR: the biodiversity cartel', in M.R. Muller and I. Lapeña (eds) *A moving target: genetic resources and options for tracking and monitoring their international flows*, IUCN, Gland, Switzerland

Vogel, J.H. (2007b) 'From the "tragedy of the commons" to the "tragedy of the commonplace": analysis and synthesis through the lens of economic theory', in C.R. McManis (ed.) *Biodiversity and the law. Intellectual property, biotechnology and traditional knowledge*, Earthscan, London

Voumard, J. (2000) *Commonwealth public inquiry. Access to biological resources in Commonwealth areas*, Commonwealth of Australia, Canberra

Wang, Y., Xiao, J., Suzek, T.O., Zhang, J., Wang, J. and Bryant, S.H. (2009) 'PubChem: a public information system for analyzing bioactivities of small molecules', *Nucleic Acids Research*, vol. 37, pp.W623–W633

Ward, O.P. and Singh, A. (2005) 'Omega-3/6 fatty acids: alternative sources of production', *Process Biochemistry*, vol. 40, pp.3627–3657

WCMC (World Conservation Monitoring Center) (1996) *The diversity of the seas: a regional approach*, World Conservation Press, Cambridge

Wegelein, F.H.T. (2005) *Marine scientific research. The operation and status of research vessels and other platforms in international law*, Martinus Nijhoff Publishers, Leiden

Weiner, R.M., Colwell, R.R., Jarman, R.N., Stein, D.C., Sommerville, C.C. and Bonar, D.B. (1985) 'Applications of biotechnology to the production, recovery and use of marine polysaccharides', *Nature Biotechnology*, vol. 3, pp.899–902

Wesche, P.L., Gaffney, D.J. and Keightley, P.D. (2004) 'DNA sequence error rates in Genbank records using the mouse genome as a reference', *DNA Sequence*, vol. 15, no. 5/6, pp.362–364

Willatts, P. and Forsyth, J.S. (2000) 'The role of long-chain polyunsaturated fatty acids in infant cognitive development', *Prostaglandins, Leukotrienes and Essential Fatty Acids*, vol. 63, no. 1/2, pp.95–100

Williams, D.H., Stone, M.J., Hauck, P.R. and Rahman, S.K. (1989) 'Why are secondary metabolites (natural products) biosynthesized?', *Journal of Natural Products*, vol. 52, pp.1189–1208

Williams, M.J., Ausubel, J., Poiner, I., Garcia, S.M., Baker, D.J., Clark, M.R., Mannix, H., Yarincik, K. and Halpin, P.N. (2010) 'Making marine life count: a new baseline for policy', *PLoS*, vol. 8, no. 10, pp.1–5

Williamson, M.H. (1997) 'Marine biodiversity in its global context' in R. Ormond, J.D. Gage and M.V. Angel (eds) *Marine biodiversity: patterns and processes*, Cambridge University Press, New York

Winter, G. (2009) 'Towards regional common pools of GRs – improving the effectiveness and justice of ABS', in E.C. Kamau and G. Winter (eds) *Genetic resources, traditional knowledge, and the law solutions for access and benefit sharing*, Earthscan, London

Wolff, N. (2004) 'Meeres- und Küstenbiodiversität', in N. Wolff and W. Köck (eds) *10 Jahre Übereinkommen über die biologische Vielfalt: Eine Zwischenbilanz*, Nomos, Baden-Baden, Germany

Wolfrum, R. (1996) 'The Convention on Biological Diversity: using state jurisdiction as a means of ensuring compliance', *Beiträge zum ausländischen öffentlichen Recht und Völkerrecht*, vol. 125, pp.373–393

Wolfrum, R. (2006) 'Hohe See und Tiefseeboden (Gebiet)', in W. Graf Vitzthum (ed.) *Handbuch des Seerechts*, Verlag C.H. Beck, Munich

Wolfrum, R. and Matz, N. (2000) 'The interplay of the United Nations Convention on the Law of the Sea and the Convention on Biological Diversity', *Max Planck Yearbook of United Nations Law*, vol. 4, pp.445–480

Wolfrum, R., Klepper, G., Stoll, P.-T. and Franck, S.L. (2001) *Genetische Ressourcen, traditionelles Wissen und geistiges Eigentum im Rahmen des Übereinkommens über die biologische Vielfalt*, BfN, Bonn

WTO (World Trade Organization) (2001) 'Doha Ministerial Statement', WTO document WT/MIN(01)/DEC/1

WTO (2006a) 'The relationship between the TRIPS Agreement and the Convention on Biological Diversity', WTO document IP/C/W/368/Rev.1

WTO (2006b) 'Review of the provisions of article 27.3(b)', WTO document IP/C/W/369/Rev.1

WTO (2008) 'Draft modalities for TRIPS related issues', WTO document TN/C/W/52

WTO (2012) 'Article 27.3b, traditional knowledge, biodiversity', www.wto.org/english/tratop_e/trips_e/art27_3b_e.htm, accessed 12 November 2012

Yin, Y. and Zheng, X. (2008) 'Marine microbial fuel cells: new technology for energy generation', *Journal of Biotechnology*, vol. 136S, pp.S593–S594

Yorktown Technologies (2012) 'GloFish. Experience the Glo', www.glofish.com, accessed 6 November 2012

Young, T.R. (2004) 'An implementation perspective on international law of genetic resources: incentive, consistency, and effective operation', *Yearbook of International Environmental Law*, vol. 15, pp.3–93

Young, T.R. (2008) 'The challenge of a new regime: the quest for certainty in access to genetic resources and benefit-sharing', *Asian Biotechnology and Development Review*, vol. 10, no. 3, pp.113–136

Zewers, K.E. (2008) 'Bright future for marine genetic resources, bleak future for settlement of ownership rights: reflections on the United Nations Law of the Sea Consultative Process on marine genetic resources', *Loyola University Chicago International Law Review*, vol. 5, no. 2, pp.151–176

Index

access and benefit sharing (ABS):
    Australia 77–90, 96–8, 105; Belgium
    99; Bulgaria 99–100; CBD-UNCLOS
    comparison 52–3, 56–9, 61–2, 71–2,
    76, 78–84; concept 21–2; European
    Union 95, 99; general CBD
    provisions 45–51, 68–70; general
    UNCLOS provisions 43–5, 66–7;
    Germany 100; Greenland 101–2;
    impairing research and development
    112, 119–22; ineffectiveness 118–19;
    in internal water and territorial sea
    52–3, 70; injustice 112–14; in the
    EEZ and continental shelf 53–9,
    70–1; in the high seas and the Area
    59–64, 71; Italy 100; Malta 100;
    Northern Territory 90–2, 96–8, 105;
    Norway 102–6; Portugal 100–1;
    problems 1–2; Queensland 92–4,
    96–8, 105
accession number 128, 139–42, 144–5,
    167, 173
actual and potential value 21, 34, 37, 180
African Model Legislation for the
    Protection of the Rights of Local
    Communities, Farmers and Breeders,
    and for the Regulation of Access to
    Biological Resources 41
agreements 26, 45–8, 58–9, 62–5,
    71, 74, 168; ABS 47, 63, 65, 74,
    76, 110–11, 175, 181; access 113;
    benefit-sharing 48, 65, 87–8, 91–4,
    96; bioprospecting 113; case studies
    106–8, 110; databases 130–2, 137,
    155–6; Data Sharing Agreement
    152–4, 157; Data Use Agreement
    151–2, 154, 157; implementation
    62–4, 117; international 27, 58, 78,
    84, 95; material transfer 46, 48–9, 65,

96; multilateral 45; national 99, 102;
    regional 166, 169
agrochemicals 4–5, 14
Andean Community 116, 166
antifoulants 5–6, 14
AquAdvantage salmon 9, 14, 141, 160
Arasena-A see spongothymidine
arbitration 72–5, 83, 95
areas beyond national jurisdiction see
    jurisdiction
areas within national jurisdiction see
    jurisdiction
ASEAN Framework Agreement on
    Access to Biological and Genetic
    Resources 116–17, 166
astaxanthin 9, 15, 159, 161–2
AstraZeneca 107–9
awareness 51, 82, 119, 156

Bancol see Bensultap
barcoding 13, 16
Basic Local Alignment Search Tool 140,
    144
beneficiary 30, 64, 68, 169
Bensultap 5, 14, 158
bias 152, 157, 173
bilateral 2, 45, 112, 114–16, 168–9, 171,
    178
bioactive molecules 6, 11, 39, 107
BioAstin see astaxanthin
biodiscovery 90–4, 96–8, 109
Biodiscovery Act 2004 (Queensland)
    90, 92, 94, 96, 109
biodiversity see biological diversity
Biodiversity Action Plan 95
biodiversity cartel theory 117
Biodiversity Strategy of the European
    Community 95
biofuels 7

biogeography 17, 122, 149, 168, 173, 176
bioinformatics 38–9, 122–4, 144
biological databases: background 2, 122–5; biochemical databases 125–38; evaluation 161, 166–76, 179; exemplary application 155–65; gene databases 138–42; protein databases 142–6; biogeographic databases 146–55
biological diversity 56, 58, 114; areas beyond national jurisdiction 61–3; Australia 93; benefit sharing 166, 169; Bonn Guidelines 46; Bulgaria 99; common concern 33; conservation 21, 33, 50–1, 63, 114, 169, 171; definition 21–2; Germany 100; global multilateral benefit sharing mechanism 51; monitoring components of 73–4; Nagoya Protocol 50; value 37
biological information 39, 110, 122, 173, 178
biological molecules 3–4, 16–18, 36, 39–40, 64, 110; biological databases 124–5, 155, 160–1, 166, 173–4, 178–9
biological resources 32, 34, 101, 106; Australia 85–96; components 21–2, 32; definition 21; utilization 37–41
Biological Resources Access Agreement Between the Commonwealth of Australia and the Craig J. Venter Institute 106–7
Biological Resources Act 2006 (Northern Territory) 90–1, 96
biological source 6, 14–15; databases 125, 128, 130, 132–4, 136, 139, 155, 160, 179
biological systems 37, 39, 110, 178
biopiracy 32, 119, 121, 178
bioprospecting 44, 91, 96, 103–4, 113, 181
bioremediation 6–7, 14, 38
biotechnology 110, 123, 127–8, 136, 142, 145; application 8, 10, 11, 123, 128; marine biotechnology 4, 17–18; potential 168; research 68–9, 80
blockbuster 176
Bonn Guidelines 21, 78–86, 100, 106; access 43, 46–50, 56; benefit sharing 68–70; monitoring 73–4; stakeholders 20

capacity 54, 57, 62, 80–1, 103; building 49, 53, 69–70, 82, 95, 110; technological 20, 28, 67, 109
Cartap 5, 14, 158

cefixime 13, 15, 158–9, 163, 165
certificate of compliance 75, 79, 82
chain of valorization *see* value chain
Charter for Economic Rights and Duties of States 114
checkpoints 74–5, 82, 103, 122
Chemical Abstract Service 129, 155
CHEMnetBase 126
chilling effect 120
coastal state 28–31, 43, 52–62, 66, 70–6, 169, 177
coastal waters 85, 90
collection authority 93, 94, 96
Combivir *see* spongouridine
common concern of humankind 33, 181
common heritage of mankind 26–34, 60, 62, 169
Community Cyberinfrastructure for Advanced Microbial Ecology Research and Analysis 107
competent national authority 47–8, 50, 78, 100, 103, 170
competition 113–14
compliance 49, 58, 65, 72–7, 79, 82–3, 100–6; code 93, 96, 97; measures 103; non-compliance 62, 74–6, 94, 118–19, 153; user 103
conotoxin 12, 15, 159, 163, 165
conservation of biological diversity *see* biological diversity
continental shelf: access 53–4, 56, 58, 60; Australia 85, 88–90, 98; benefit sharing 70; coverage 45; monitoring and compliance 76; Nagoya Protocol 116; Norway 104; powers 31, 78, 177; zonation 24–5
CBD Convention on Biological Diversity: Ad Hoc Open-Ended Working Group on Access and Benefit Sharing 46, 49; Clearing-House Mechanism 47, 75, 82, 95, 170; Conference of the Parties 22, 46, 50, 75; jurisdictional scope 61, 66; Precedence to UNCLOS 56, 58–9, 110; shared natural resources 116
countries with occurrences 151, 153–4, 160
country of origin 47, 50, 100, 102–3, 138, 156, 176
CRC Press 126
CRCnetBASE 126
Cytosar-U *see* spongouridine

declaration of the origin 178

Declaration of the United Nations Conference on the Human Environment 114
derivative 64–6
DHASCO *see* docosahexaenoic acid
Dictionary of Marine Natural Products 124–30, 156, 158, 172
dispute settlement 50, 72–7, 79, 83
distribution of marine species 16–17
DNA Databank of Japan 139
DNA polymerase 8, 14, 38, 145, 158, 162, 164
docosahexaenoic acid 9, 15, 159–60, 162, 164
dualist state 22

EC Biodiversity Clearing House Mechanism 95
EC Directive on the Legal Protection of Biotechnological Inventions 95, 100
economic development 20, 22, 26, 28, 67
ecteinascidin 12, 15, 130–1, 133, 139, 159, 163, 165
effectiveness 2, 118, 122, 161, 166, 173, 178–9
enforcement 83, 118; Australia 88, 93–4, 96–7, 105–6; benefits 69–70; dispute 75
environmental law 27–8, 114
Environment Protection and Biodiversity Conservation Act 1999 (Australia) 85–6, 88, 90, 96; Regulations 86–8, 90, 96
European Bioinformatics Institute 144
European Nucleotide Sequence Database 139
Evisect S *see* Thiocyclam
Exclusive Economic Zone 78, 80, 83–4, 147; access 53–4, 56, 58–60; Australia 85, 88–9; Belgium 99; benefit sharing 70; CBD coverage 45; compliance 76; Denmark 102; dispute 73; Nagoya Protocol 116; Norway 104; Portugal 101; powers 30–1, 76, 177; zonation 24

facilitation: access to genetic resources 22, 46–7, 53, 56–7, 78, 85; access to knowledge and technology 67–8; non-commercial research 44, 51, 55, 57, 71–2, 83
Fisheries Act 1988 (Northern Territory) 91–2, 94, 98

Food and Agriculture Organization 31, 115
Foreign Research Vessel Guidelines 1996 (Australia) 89–90
Framework Agreement of the Hindu Kush-Himalayan Region 117, 166
freedom of the high seas *see* high seas
functional units of heredity 21, 34–7, 39, 42, 64–5, 180–1
fund 115, 117, 166, 175, 178

GenBank 124, 138–44, 147, 155–6, 160–1, 167
gene cooperatives 117
genetic expression 35–7, 64, 83
genetic material 38, 77, 102–4; definition 21, 34, 36–7
genetic resources 27, 41–2; conflicts from utilization 20–1; definition 1, 21, 26, 34–7, 49; legal status 1, 26, 30–4
German Research Foundation 100
Global Biodiversity Information Facility 124, 147, 150–5, 157, 160, 162, 166–7, 176
Global Environment Facility 170–1
global multilateral benefit sharing mechanism 51, 62, 84, 168–72, 175; biological databases 161, 169, 179
GloFish *see* green fluorescent protein
good neighbourliness 114
grand global bargain 168
green fluorescent protein 13, 15, 160, 163, 165
Griffith University Eskitis Institute for Cell and Molecular Therapies 107–9

halogenated (brominated) furanones 6, 14, 158, 160, 162, 164
high seas 28, 45, 56, 59–63, 169, 177; zonation 24–5
Human Genome Project 35–6

implementation agreement *see* agreements
indigenous communities *see* traditional knowledge
industrial-enzymes 8
industry 1, 5, 10, 12, 17–18, 42, 102
inequitable use 113–14
Informa 126
injustice *see* access and benefit sharing
intellectual property rights 32, 44, 47, 49, 150; biopiracy 119; databases

132, 135–6, 138, 142, 155, 161,
164–7; patents 74, 80, 84, 95,
99–105, 175, 179
Intergovernmental Committee for the
Nagoya Protocol 75
Intergovernmental Oceanographic
Commission 55, 67, 147
intermediaries 19–20, 66, 118, 169, 175
internal waters 116; access 45, 52–3;
Norway 104; powers 30, 177;
zonation 24
International Court of Justice 73–4, 78,
114
international instruments 1–2, 18, 21,
24, 28, 42, 178
international organization 101, 152, 177;
UNCLOS 55, 61, 67, 78–9, 83
international regimes 2, 33, 49, 177
International Seabed Authority (ISA)
31, 60–3
International Treaty on Plant Genetic
Resources for Food and Agriculture
32–3, 115, 170
International Tribunal for the Law of the
Sea 73, 78
International Undertaking on Plant
Genetic Resources 31–3
isothiazolon 6, 14, 158, 162, 164

JBR425 7, 14, 158
J. Craig Venter Institute 106, 110
junk DNA 36
jurisdiction 19, 54, 57, 59, 75, 77,
104, 106; areas beyond national
jurisdiction 23, 25, 28–32, 61–6,
168–70; areas within national
jurisdiction 30, 33–4, 45, 110,
168–70; Australia 85–6; Belgium 99;
Europe 95, Greenland 102

Kansas Geological Survey Mapper 147,
149

Life'sDHA *see* docosahexaenoic acid

mariculture 8–9, 14, 18
marine environment 3–4, 104–5, 150,
152, 169, 177; legislation 1, 22–3,
28, 30, 43–4, 55, 58, 67, 73; Norway
104; products 6–9, 17–18; sovereign
rights 30–1; species distribution
16–17
marine genetic resources: common
heritage of mankind 29; coverage
by the CBD 22, 45–51; coverage
by UNCLOS 23, 42–4; derivatives
see derivative; problems 1–2, 18;
sovereign rights 27; stakeholders
19, 175–6; uses 3–16; utilization *see*
utilization of genetic resources
marine scientific research: applied
research 44; Australia 89–90; basic
research 44; benefit sharing 67,
71–2; common heritage of mankind
31; compliance 76; definition 43–4;
derivatives *see* derivative; dispute
settlement 73; EEZ and continental
shelf 54–9; general UNCLOS
provisions 23, 44–5; high seas and
the Area 60–2; internal waters and
territorial sea 52; marine genetic
resources 53, 59, 62, 105; Norway
104; Portugal 101
maritime zones 23–5
material transfer agreements *see*
agreements
memorandum of understanding 152–3
Merck & Co Inc. 132, 134
Merck Index 124, 131–4, 155–6, 158–9,
164–5, 172
monist state 22
monitoring: 47, 61, 64, 105–6, 117–22;
Australia 88, 93–4, 96–7; CBD and
UNCLOS 72–7, 80, 82; biological
databases 124, 127, 132, 135, 145,
161, 166–76, 179
multilateral 114, 116
Multilateral System of the Food and
Agriculture Organization 115–16,
170
mutually agreed terms 78–9, 96, 100,
105, 113; access 45–50, 53, 56,
68, 77; benefit sharing 68, 70,
181; databases 142, 153, 167, 176;
monitoring 74–5, 137; Multilateral
System 115

Nagoya Protocol 1, 21, 78–84, 106, 110;
access 43, 50–3, 57; benefit sharing
68, 70; databases 122, 150, 161, 166,
168–72, 175, 179; derivative 35;
genetic resources beyond national
jurisdiction 62–4, 177; monitoring
73–5; transboundary cooperation
116; utilization of genetic resources
37, 41, 178
NAPRALERT *see* Natural Products
Alert

National Center for Biotechnology Information (NCBI) 124, 136–45, 142, 155–6, 158–9, 161, 164, 172, 176
national focal point 47, 49–50, 75, 80–2, 95
National Institutes of Health 137
National Library of Medicine 137, 142–3
Natural Products Alert 124, 134–6, 155–9, 164–5
NatureBank 108
NCBI Protein Database 124, 142–3, 155–9, 161
nereistoxin 5, 14, 158, 162, 164
Netsafe 6, 14, 158
neutraceuticals 9–10

Ocean Biogeographic Information System 124, 140, 146–50, 152, 155, 157, 182
opAFP-GHc2 *see* AquAdvantage salmon
Organization for Economic Co-operation and Development 4, 38, 41
outdated 152, 155, 173
ownership 27, 69, 87, 91, 97, 108, 116, 119; databases 134, 150

Padan *see* Cartap
Pardo, A. 28–9
party providing genetic resources 2, 20–2, 41, 45, 47, 105–6; access 48, 53, 57–9, 65, 112, 114, 118–22; benefit sharing 68, 70, 72, 76, 109–10, 178–9; biodiversity cartel 117; databases 122, 167, 170–6, 179; definition 19; interests 20; Norway 103; transboundary situation 170
patent *see* intellectual property rights
Pearlsafe 6, 14, 158
permits 50, 74–5, 79, 86–96, 101, 103–5
persistent identifiers 167–8, 174
personal care 10
pharmaceuticals 11–13, 19, 38, 42; ABS partnerships 107–8; databases 131–2, 135
plant genetic resources 32, 115
precautionary principle 88
Prialt *see* conotoxin
prior informed consent 45–51, 56, 63, 65, 68, 74–7; Australia 91; databases 142, 153, 176; global multilateral benefit sharing mechanism 168–9; Multilateral System 115
Protein Information Resource 144

provider 17, 19–20, 46, 65, 75; Australia 87–92, 105; measure 104, 111–12, 118–19, 121–2, 178; Norway 102; Portugal 101; state *see* party providing genetic resources
pseudopterosin A 10, 15, 159, 162, 164
PubChem 124, 136–40, 142, 155–6, 158–61, 172
public domain 107, 138, 142–5, 172
PubMed 136–9, 142–3, 145, 161, 167, 172

Queensland Herbarium 108–9
Queensland Museum 108–9

red fluorescent protein 13, 15, 160, 163, 165
regional fisheries organizations 62–3
remedies 74, 83, 177
reporter genes 13
res communis 60, 78, 177
research and development 2–7, 17–19, 84, 178–80, 166–7, 180; Australia 86, 98, 105, 108; benefit sharing 66, 68–70, 79–80; databases 161, 174–5; hampered 119–22; monitoring 172; pharmaceuticals 11; prior informed consent 48; utilization of genetic resources 37, 42, 110, 115, 118, 178
research community 120, 174
research state 23, 55, 59, 67, 70–3, 76, 79
Resilience *see* pseudopterosin A
rhamnolipid 6–7, 14, 158, 162, 164
Rio Declaration on Environment and Development 31, 114

Sea-Nine 6, 14, 158
Seas and Submerged Lands Act 1973 (Australia) 88, 90
self-determination 26–7, 53
shared natural resources 114–15
sovereign rights over natural resources 22–3, 26–33, 46, 50, 66, 110; comparison with common heritage 29–30
sovereignty 17, 32, 51–3, 57, 70, 116, 177; over natural resources 27–31, 33
Sorcerer II Global Ocean Sampling Expedition *see* J. Craig Venter Institute
source states *see* countries of origin

Spirulina Pacifica 9, 15, 158, 162
spongothymidine 12, 15, 128–9, 159,
  162, 164
spongouridine 12, 15, 159, 162, 164
superoxide dismutase 10, 15, 158, 162
Suprax *see* cefixime
surplus of allowable catch 31, 54, 78
sustainable use 120; Australia 86, 91,
  96; Bulgaria 99; CBD aim 21, 41,
  46, 50–1, 56, 58, 78, 116; databases
  149, 166, 171–2; global multilateral
  benefit sharing mechanism 168–9;
  ITPGRFA 115; monitoring 73–4;
  Norway 113; relationship to ABS 70
Swiss Institute of Bioinformatics 122,
  144
Swiss-Prot 142, 145

taxonomy 13, 16, 18, 46, 173, 176;
  Australia 87, 109; benefit sharing
  166; Bonn Guidelines 46; Nagoya
  Protocol 41–2, 48; databases 123,
  140, 143–8, 150–2, 168, 182
Taylor and Francis 126
territorial sea 24–5, 30, 45, 52–3, 56,
  70–1; Australia 89, 98; Norway
  103–4; territory 116
the Area 24–5, 29, 31, 59–64, 70–1, 169,
  177
Thieme 129–30
Thieme RÖMPP Online 124, 127–31,
  155–6, 158–9, 161, 164–5
thiocyclam 5, 14, 158
third parties: Australia 107–8;
  biodiversity cartel 117; databases
  153; Greenland 102; monitoring 173;
  Norway 104; transfer 47, 65–6, 84,
  87, 91, 99, 120
tracing *see* monitoring
traditional knowledge 46–7, 50–1, 73–8,
  86, 105; Australia 96; benefit-sharing
  agreement 91, Portugal 101
Trade Related Aspects of Intellectual
  Property Agreement 106

transaction: ABS 2, 68, 112, 115–16,
  120–1, 168–78; cost 48, 117–18, 120,
  170
transboundary: conflicts 113; cooperation
  51, 84, 116–17, 166, 172, 175, 179;
  genetic resources 63, 122; situations
  51, 84, 115, 168–70
transfer of technology 43, 49, 66–72, 80,
  84
Trizivir *see* spongouridine

UniProt *see* Universal Protein Resource
United Nations Charter 26–7, 72
United Nations Convention on the Law of
  the Sea 1, 21–6, 30–1, 42–5, 110–11,
  178
United Nations Environment Programme
  Environmental Law Guidelines
  and Principles on Shared Natural
  Resources and Principles 114
United Nations Fish Stocks Agreement
  115
United Nations General Assembly 26–31
United Nations Security Council 27
Universal Protein Resource 124, 143–6,
  155, 157, 159, 164–5
University of Illinois 135
user: behaviour 105–6, 178; measure 95,
  99–100, 103–5, 111, 119, 178; pioneer
  167, 173, 178; state 20–1, 47, 68, 74,
  77, 118–23, 171, 177–9
utilization of genetic resources 37–42;
  benefits 68–72

valorization *see* value chain
value chain 2, 112, 118, 121
Venuceane *see* superoxide dismutase

World Summit on Sustainable
  Development 49

Yondelis *see* ecteinascidin

Zovirax *see* spongothymidine